Orbital debris from upper-
stage breakup

Orbital Debris from Upper-Stage Breakup

Edited by
Joseph P. Loftus Jr.
NASA Johnson Space Center
Houston, Texas

Volume 121
PROGRESS IN ASTRONAUTICS AND AERONAUTICS

Martin Summerfield, Series Editor-in-Chief
Princeton Combustion Research Laboratories, Inc.
Monmouth Junction, New Jersey

Published by the American Institute of Aeronautics and Astronautics, Inc.
370 L'Enfant Promenade, SW, Washington, DC 20024-2518

American Institute of Aeronautics and Astronautics
Washington, D.C.

Library of Congress Cataloging in Publication Data

Orbital Debris from Upper-Stage Breakup/edited by Joseph P. Loftus Jr.
p. cm.—(Progress in astronautics and aeronautics; v.121)
1. Space debris.
I. Loftus, Joseph P. II. Series.
TL1489.072 1989 vol. 121 629.4—dc20 89-36977
ISBN 0-930403-58-4

Table of Contents

Control of Debris from Breakups

Prevention of Breakups: Upper-Stage Operational Procedures

Preface

At NASA Johnson Space Center a conference was held May 14–15, 1987, to discuss upper-stage breakups and their contribution to the Earth-orbital-debris environment and related design and operational issues. Participants included launch-system operators from NASA and Western Europe who have experienced such events, and organizations that generate data pertaining to breakups.

The presentations focused on the following questions:

What do we know?
How well do we know it?
What are the options for preventive and ameliorative action?
What are the designs and operational practices in use?
Are these designs and practices adequate or should they be modified?

The discussions in summary reflect a consensus on a number of issues:

1) Low Earth orbit (LEO)—i.e., between 500- and 2000-km altitude—contains a growing mass of man-made objects and products of breakup events. This mass has been increasing over the past 25 years, and now dictates the design environment for future low Earth orbit operations.

Since space operations began, some 18,000 objects have been cataloged as injected into space, almost 7000 of which are still in current records. During the days and weeks following the inadvertent breakup of an upper stage or the intended destruction of a spacecraft, objects detected in Earth orbit can increase by very significant factors. The persistence of these objects in Earth orbital space is a function of the orbital parameters of the parent body.

2) Historically, the largest uncontrolled contribution to new orbital objects has been the breakup of upper-stages. Upper-stage breakups have occurred for reasons that are poorly understood. The dominant cause in rocket bodies appears to be pressure-vessel failure due to deflagration of hypergolic propellants, stress failures of the vessels, or "hybrid events" such as the reduction of pressure-vessel integrity by impingement of natural meteoroids or small particles of man-made debris on otherwise intact systems.

3) The potential hazard from particles resulting from breakup of man-made objects is essentially indeterminate as to magnitude, although it is obviously large and growing. It is indeterminate because of the limits of current detection systems—radar, IR, and optical—and our knowledge of the properties of the systems on orbit, both before and after breakup.

We do know that the radar-detectable population in LEO from 500- to 2000-km is growing; and we do know that optical and IR data indicate that the radar data underestimate the total number of objects by a large factor, how large we cannot assess because neither radar, optical, nor IR data can provide assessment of size without many assumptions as to intervening variables that are not directly measureable (e.g., reflectivity, albedo, rotation velocity, and material observed) and because of the inherent properties of the detection systems themselves (e.g., as the size of objects of interest falls below the basic wavelength of the radar, detection, while possible, becomes statistically less certain, as detection occurs only due to Rayleigh scattering). Fortunately, for the smaller particles it is most important to know the statistical values so that passive protection measures can be designed; orbital elements are not required.

4) A number of practical strategies can reduce the potential for upper-stage breakups.

Stages that must be left on orbit because of current and projected performance commitments can be made significantly more inert by operational techniques. Since 1981, when it was determined that a number of Delta stages had broken up as long as three years after mission termination, revised procedures that deplete propellants and pressurants have virtually precluded such events.

ESA is considering the adoption of more positive propellent and pressurant depletion procedures for Ariane 3 and 4, particularly for missions to Sun-synchronous orbits. Design studies have been initiated for Ariane 5 to assess a controlled deorbit to ocean disposition for LEO missions. Sooner or later, all operators must do so.

5) Geosynchronous transfer stages (GTO) represent a case of particular interest. Unlike Sun (helio)-synchronous and some other LEO stages, they do not stay in nearly the same orbit as the payload they deliver. Two cases are of interest: A stage subject to autogenous breakup, in which case it will produce a long-term hazard to LEO and an increased GSO hazard; and hazard of cross contamination between LEO and GSO by spacecraft in elliptical transfer orbits, i.e., a transfer stage may strike another object while traversing LEO and thus carry high-velocity debris through GSO.

GSO is inherently benign. Man-made objects in GSO may eventually collide, but at velocities in the order of less than 800 m/s. Such events may destroy economic utility, but should not create large-scale environmental effects like the high-velocity (7,000–10,000 m/s) collisions in LEO.

6) In essence, we are endeavoring to discriminate between two classes of events—explosions and hypervelocity collisions—and the conditions that contribute to their frequency.

Explosions can be of low or high intensity. The number of pieces of debris will be a function of the energy level of the explosion.

Of critical concern are hypervelocity collisions, particularly involving high-density objects, e.g., a typical spacecraft. A hypervelocity collision will exceed the speed of sound in the structural material of the target at velocities greater than 4 km/s. At such velocities, the target will respond elastically like a fluid creating an extremely large number of small particles. At impact velocities less than 2 km/s, materials deform more in accord with normal experience of mechanical deformation; there may be some shattering and small-particle formation, but it will be comparatively very limited.

Rocket bodies as "largely empty shells" have more benign properties under impact than do dense bodies such as current spacecraft.

GTO stages can be disposed of for little velocity penalty if operational life is extended—a very different case from the disposition of heliosynchronous or other LEO stages, which require a significant velocity change or payload penalty.

7) All agree: any stage left in long-duration Earth orbit should be made inert to the maximum degree feasible—i.e., all working pressurant gas and stored propellant should be reduced to the minimum.

When design and operational procedures allow, it is proper to move to controlled ocean deposition of stages; but the timing of such action requires international agreement so as not to bias competitive performance.

GSO transfer stages are not so critical as Sun-synchronous and other LEO stages. It is desirable to minimize orbital duration primarily to preclude cross contamination from the LEO to GSO environment. Controlled venting of propellents and pressurants could significantly reduce the orbital lifetime of GSO transfer stages. Any action to regulate such stages should be later than and consistent with controls imposed on LEO stages.

8) A sensor is required that can continuously monitor the number of small debris particles (order of 1-cm diam) as a function of altitude in LEO. This sensor should provide statistical data needed for spacecraft design, plus data not currently available as to the occurrence and severity of unexpected breakup events.

9) Greater international cooperation among space-faring nations will be necessary to better monitor and assess the debris environment and to develop effective measures to prevent proliferation of orbital debris.

Orbital debris is an issue of significance to each nation in proportion to the number of launches and the operational life of each of its asset.

In conclusion, the participants agreed on the following steps:

- Each participant would do what was practical to define the environment and improve the small-particle definition of the environment.
- As soon as possible, each fluid-stage operator would effect operating procedures that would make heliosynchronous and other LEO stages as inert as possible.
- GTO-stage-management options would be studied intensively with the objective of minimizing future hazards.

However, the holdings of the participants do not necessarily reflect the official policies of the organizations represented.

Joseph P. Loftus Jr.
Assistant Director
NASA Johnson Space Center
Conference Chairman

Background

Chapter 1

Current Orbital Debris Environment

Donald J. Kessler*
NASA Johnson Space Center, Houston, Texas

Introduction

Orbital debris was first studied in the early 1970's by NASA Johnson Space Center and NASA Marshall Space Flight Center. These early studies concentrated only on the less than 2000 objects that were cataloged by North American Aerospace Defense Command (NORAD) at that time. The major conclusion of these studies was that the collision probabilities, although of some concern, were not significant except for large structures (greater than 100 m in diameter). In the mid-1970's NASA Langley Research Center expanded the level of concern by predicting that, as a consequence of explosions in space, the real orbiting population was 2.5 times the cataloged population.[1] In the late 1970's studies at the Johnson Space Center concluded that fragments resulting from collision between artificial satellites would be a major source of debris, possibly as early as the 1990's.[2] Later studies identified explosions of the U.S. Delta rocket second stage and USSR satellite tests as major contributors to debris in space.[3] As a result, the United States changed the Delta rocket operational procedures to minimize the possibility of future explosions. In 1981 the American Institute of Aeronautics and Astronautics (AIAA) issued a position paper on space debris, concluding that the space debris issue is real and that action must begin now to forestall a serious problem in the future.[4]

In 1982 the Johnson Space Center sponsored a workshop on orbital debris.[5] Solid rocket motors fired in space were identified as a potential source of very small orbital debris; some experiments flown in Earth orbit to detect meteoroids actually detected mostly aluminum oxide particulates from these rockets. Therefore, in the two debris size ranges that were measured (the micrometer size detected by meteoroid sensors and the greater-than-10-cm size detected by ground radar), the debris flux in certain regions of Earth orbit was either comparable to or greatly exceeded the interplanetary meteoroid flux. The workshop concluded that orbital debris

*Orbital Debris Project Scientist, Space Science Branch.

is a potential problem for future space operations, but that, before initiating major efforts to control the environment, new data should be obtained on the population smaller than 10 cm. Consequently, since the workshop the major emphasis has been upon obtaining such data. Most of this work, as summarized by Kessler,[6] has been updated here.

Fundamentals of Orbital-Debris Issues

Orbital debris is becoming a concern due to a combination of fundamental issues. First, the volume for Earth-orbiting objects is much smaller than that occupied by interplanetary meteoroids. Consequently, practically since the beginning of the space program, collision probabilities from payloads and rocket bodies alone have been orders of magnitude higher than collision probabilities from meteoroids of the same size. Second, just as in the case of interplanetary meteoroids, large objects act as a source of smaller objects by fragmenting. Fragmentation of man-made satellites occurs either by accidental explosion of rocket bodies or by collision between two objects. To date, 90 satellites are known to have broken up in Earth orbit. Moreover, in a significant number of "anomalous events" a few pieces have been observed to originate from a larger satellite. Depending on the size distribution of fragments, the mass in a single, average man-made satellite can cause the flux from these fragments to exceed the interplanetary meteoroid flux over the entire size spectrum of fragments and over several hundred kilometers of altitude. Finally, the large orbital inclinations coupled with the rapid precession of the ascending node produce high collision velocities in low Earth orbit. The average collision velocity between two objects in low Earth orbit is 10 km/s, compared to an impact velocity of 15–20 km/s for interplanetary meteoroids. Consequently, the damage caused by an orbital debris impact is not very different from that caused by meteoroids.

Sources of Orbital Debris

Cataloged Population

So far, only a few countries and the European Space Agency (ESA) can launch objects into Earth orbit. Over that last four years (Table 1) the USSR has the largest launch rate, representing about 80% of the world's launch schedule. However, since most of the USSR satellites have very short

Table 1 Launch operations

	Number of Launches in Given Year			
Country or organization	1984	1985	1986	1987
USSR	97	98	91	95
United States	22	17	6	8
Japan	3	2	2	3
China	3	1	2	2
European Space Agency (ESA)	4	3	2	2
Totals	129	121	103	110

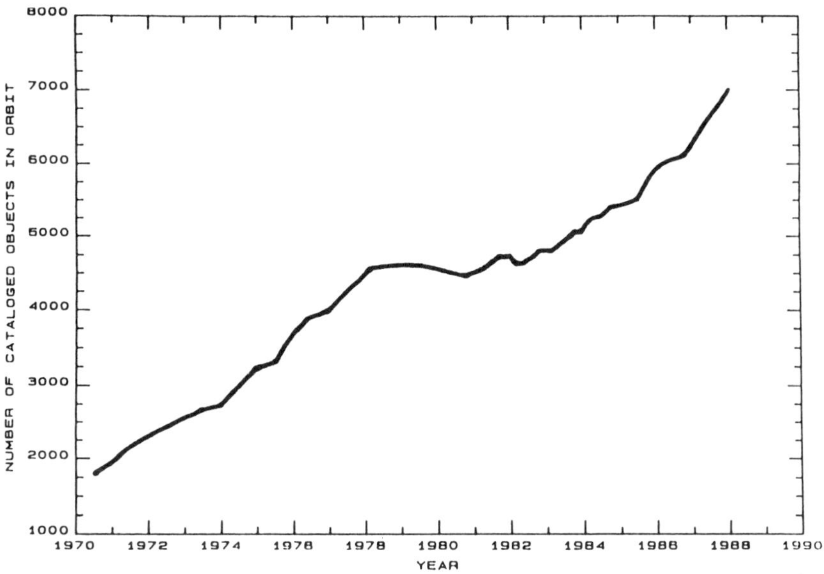

Fig. 1 Historical record of objects in space.

orbital lifetimes, their contribution to debris is less than that suggested by Table 1. Both the United States and the USSR share roughly equal responsibility for the current orbital-debris environment.

Figure 1 shows how the cataloged population has increased over the past 18 years, leading to an official "box score" of 7023 objects in orbit as of December 31, 1987.[7] The reduction in the rate of increase of objects in space between 1978 and 1983 reflects two factors:

1) This period corresponds to the second-highest solar activity since records have been kept. High solar activity heats the upper atmosphere and increases the atmospheric density by more than an order of magnitude at most satellite altitudes. This causes an above-average number of objects to re-enter the atmosphere.

2) Between 1978 and 1981 there happen to have been no major explosions. In January 1981, however, a Delta second stage exploded, and in July 1981 Cosmos 1275 broke up. These two events account for the temporary increase in number during 1981. Since 1981 all Delta stages perform a depletion burn to eliminate excess fuel after placing payload in orbit. Consequently, none of the rockets have since exploded, and future explosions are highly improbable. The Cosmos 1275 breakup is unexplained. It may be significant that this payload was in an orbit that had a relatively high probability of collision with other objects orbiting at an altitude between 900 and 1000 km.

The importance of satellite breakups is shown in Table 2. Nearly half of the cataloged population comes from satellite breakups. All but a few of these were explosions—most were accidental, but some were intentional.

Table 2 Source of in-orbit population cataloged by NORAD from December 31, 1987 population of 7023 objects[a]

Space object	Percentage of tracked population in orbit
Operational payloads	5
Nonoperational payloads	20
Mission-related (rocket bodies, shrouds, etc.)	25
Satellite breakups (explosions and unexplained)	50

[a]Distributions are roughly equally divided between the USSR and the United States. The 26 breakups account for 94% of the tracked fragments resulting from the 91 known explosions to date: 11 U.S. upper stages, 1 U.S. test, 3 USSR unknown, 10 USSR tests, and 1 ESA upper stage.

With only 5% of the catalog consisting of operational payloads, it is obvious that debris control has little to do with limiting the number of operational payloads in orbit.

Uncataloged Population

The first suggestion of a significant uncataloged population comes from a comparison of the fragment distribution from ground explosions with that detected by NORAD radars.[1,3] Based on such considerations, it has been estimated that between 10,000 and 15,000 objects larger than 4 cm are in Earth orbit. Figure 2 gives an estimate of NORAD's ability to detect objects in Earth orbit. However, an object, although detectable, may not appear in the catalog, either because of a low probability of detection or because of an administrative decision to ignore the object. As reported by Johnson, special tests conducted by NORAD radars confirm the existence of a significant uncataloged population.[8]

Explosions have the potential, however, of producing a much larger number of smaller objects. This is especially true in high-intensity explosions or in explosions where the payload is designed to break into some particular size. It is possible for a single 100-kg payload to break into 10^5 1-cm objects or into 10^8 1-mm objects.

Solid propellant rocket motors fired in space produce mostly very small particles. A single rocket can be responsible for placing 10^{20} particles in space, ranging in size from 1×10^{-4} mm to 1×10^{-2} mm. Fortunately, most of these particles reenter quickly. However, the small fraction that remains in orbit can produce a flux that exceeds the meteoroid flux, as shown by Mueller and Kessler.[9]

Unmodeled sources of small debris surely exist. The most likely sources include the material that may be released as a by-product of separation techniques used in orbit. Another possibility is shedding of spacecraft surfaces resulting from fatigue in space. For example, the thermal expansion and contraction of spacecraft surfaces may cause paint to flake off slowly. We do not have enough data to model these potential sources adequately, but new data from recovered spacecraft, as discussed later, are becoming available.

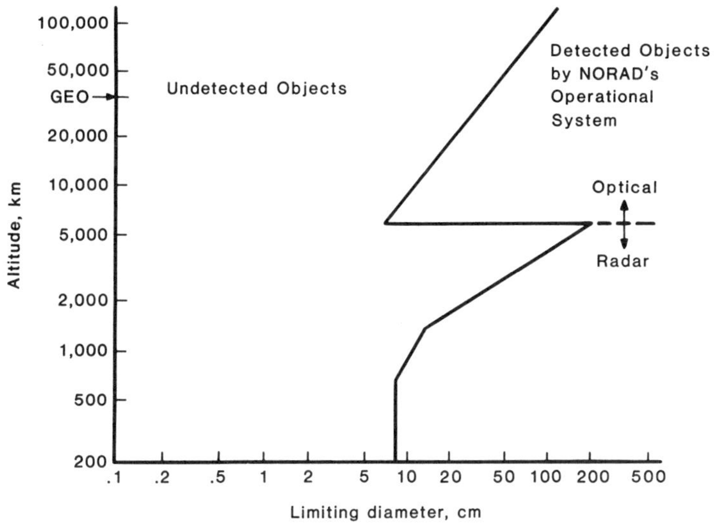

Fig. 2 NORAD's operational ability to detect objects.

Even if all of these sources were removed, a significant source would remain that is essentially a by-product of having objects in orbit: collisional fragmentation. As shown in Fig. 3, if past growth rates continue, collisions between objects larger than 4 cm can be expected within the next few years. A typical collision would be between an old rocket body or payload and an explosion fragment larger than 4 cm and untracked by NORAD.

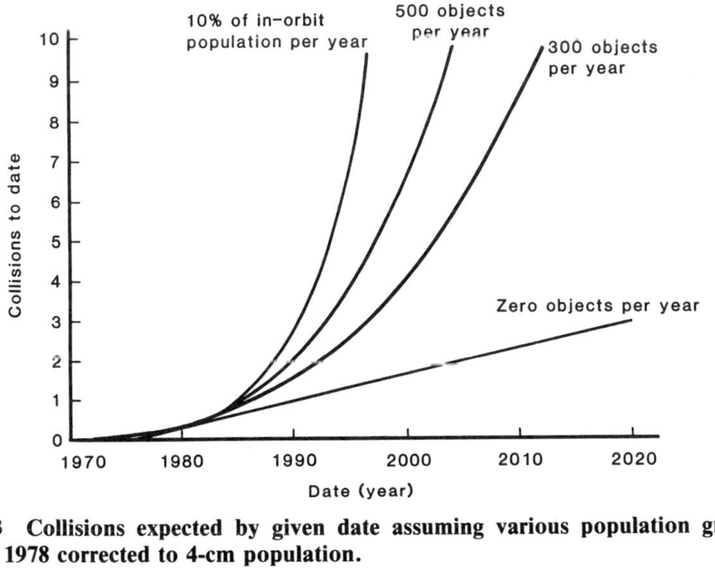

Fig. 3 Collisions expected by given date assuming various population growths from 1978 corrected to 4-cm population.

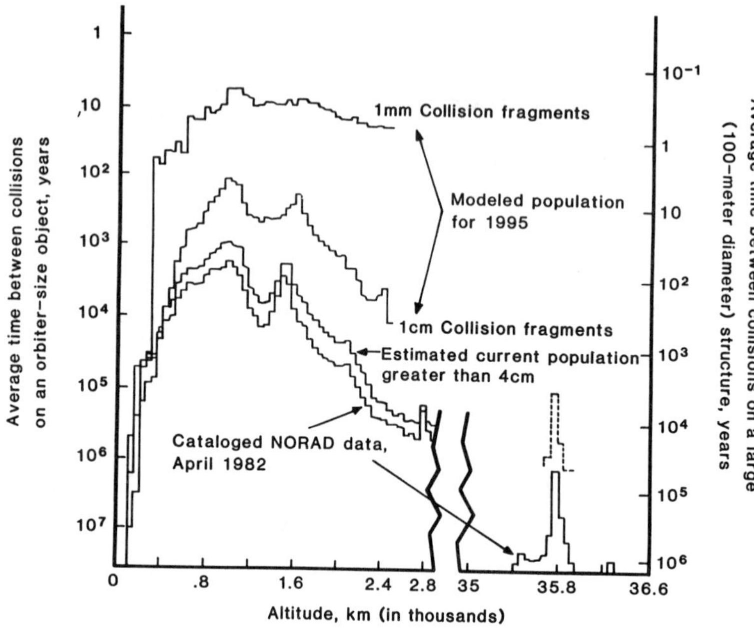

Fig. 4 Average time between collisions; modeled population for 1995 compared to current population.

The collision would produce 10^4 particles larger than 1 cm and more than 10^6 particles larger than 1 mm. Figure 4 compares the predicted 1995 environment for these two sizes resulting from collisional fragmentation with the current environment. This figure presents the environment expected in 1995, assuming that the major source of smaller particles at lower altitudes is from collisions at higher altitudes, where the fragments either drift down because of atmospheric drag or eject into lower orbits as a result of the change in velocity of collision fragments.[8] Monte Carlo modeling presented by Su and Kessler shows the results of using less change in velocities for smaller fragments.[10]

Measurements

Measurements of orbital debris at sizes smaller than those cataloged by NORAD were nonexistent until a few years ago. The first measurements of orbital debris were apparently taken in 1973-74 during the Skylab program. The Skylab Cosmic Dust Experiment S149 was returned to Earth for analysis. Chemical analysis of residues in individual craters showed a high incidence of aluminum.[11] The windows on the returned Skylab IV Apollo Command Module were also examined for meteoroid impacts using the scanning electron microscope. It was discovered that about half of the hypervelocity pits were aluminum-lined.[12] The size distribution and frequency of these pits were about what one would expect as a result of solid-propellant rocket motors being fired in space.

With the beginning of the Shuttle Orbiter launches, windows were again examined for hypervelocity impacts. However, because reflown, the Orbiter windows could not be used in destructive testing, unless they contained a pit too large for safe reuse. On the morning of the third day of the seventh Shuttle flight (STS-7), the crew detected a pit on the outside window that was about 5 mm in diameter. They were reasonably sure the pit had not been there before. This pit caused the window to be replaced, and thus could be examined. The pit contained titanium oxide with traces of carbon, aluminum, potassium, and zinc—elements consistent with those found in thermal paints using an organic binder. It was concluded that the particle has a diameter of about 0.2 mm and an impact velocity between 3 and 6 km/s.

The Explorer 46 Meteoroid Bumper Experiment was flown in 1972 to test the effectiveness of bumpers against meteoroids. The very strong directional flux measured by this experiment can only be explained by impacts from an Earth-orbiting population.[13] The experiment was sensitive to impacts by particles larger than about 0.1 mm. The orbital-debris flux determined by 43 highly directional impacts was 1.9 impacts/m^2/year. This is about a factor of 3 below the flux derived from the Orbiter window (assuming one impact every 50 days in space). The Orbiter window flux is about the same as the meteoroid flux.

The Solar Maximum Mission (SMM) satellite was launched on February 14, 1980 into a nearly circular orbit at 570-km altitude. By April 10, 1984 the orbit decayed to 500 km and SMM was captured for repair in the Shuttle payload bay, after nearly 50 months of exposure to space. The returned surfaces included about 1.5 m^2 of thermal insulation and 1.0 m^2 of aluminum thermal-control louvers. The thermal insulation consisted of 17 layers of aluminized kapton or mylar, each separated by a dacron net, and the louvers consisted of two layers of heavy aluminum foil separated by about 3 mm. The surfaces offered an excellent opportunity to obtain the chemistry of impacting particles. Analysis of these surfaces is continuing, but three different types of hypervelocity impacts have already been found: meteoritic material, paint particles (consisting both of titanium and zinc pigments, with potassium silicate binders, and of an unknown chlorine source), and particles of an unknown origin (aluminum droplets were detected, but these may have originated from the aluminized surfaces).[14] Further analyses of these surfaces should provide a more definitive definition of the debris environment for sizes smaller than about 0.2 mm.

Early studies within NASA determined that ground-based telescopes could be used to detect 1-cm orbiting debris. NASA contracted with Lincoln Laboratories to use its two 31-in. telescopes to search for this debris. The two telescopes, located 60 m apart, were pointed near the zenith just after sunset and just before sunrise. Orbiting objects were then detected by reflected sunlight transiting the field of view. The data were recorded on videotape. Data-reduction technique included applying the parallax observed between the two telescopes to discriminate against meteors that were detected near 100-km altitude. The results of this search revealed a population eight times the cataloged population, or about 40,000 objects that are 1 cm or larger.[15] These measurements are summarized in Fig. 5

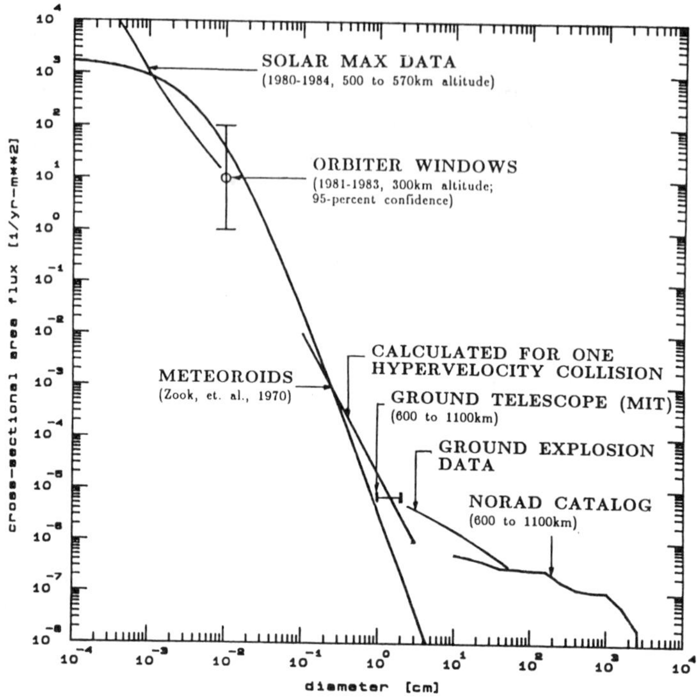

Fig. 5 Orbital debris vs meteroid flux.

and compared with the meteoroid environment.[16] In all cases the measured debris environment is comparable to, or exceeds, the meteoroid environment.

Proposed Measurements

Debris larger than 1 cm can be detected from the ground. Smaller debris must be measured in space. Debris smaller than 0.1 mm can be detected using impact sensors, as were used to detect meteoroids in space. However, those experiments must be modified to distinguish between orbital debris and meteoroids. Composition, speed, and direction are all discriminators that can be applied. Experiments of this type have been proposed by Grun[5] and Kuczera et al.[17]

Debris between 0.1 mm and 1 cm are more difficult, however, to measure. Studies within NASA have concluded that an orbiting radar, a lidar, or an optical telescope could detect this size from orbit and obtain sufficient information on trajectory to distinguish between meteoroids and debris. Sanguinet[18] and Reynolds[19] summarized many of the considerations required for such a detector. Meteoroid experiments on the Long-Duration Exposure Facility (LDEF) should also provide data.

Space-Debris Assessment Program

NASA has prepared a plan which defines the activities and resources required over the next ten years to develop the necessary understanding of the man-made orbital debris environment, its effects upon future mission operations, and the related requirements for appropriate policy and policy implementation. The major elements of the plan are environment definition, hazard assessment, space-object management, and policy. The most immediate requirement of the plan is to define the environment, with primary emphasis on collecting data on debris sizes smaller than 4 cm.

Environment definition combines data from modeling, ground tests, ground data collection, and flight data collection to define the debris environment in terms of flux and velocity as a function of altitude, orbit, debris size, the future population growth. Most of the current data have been compiled and modeled. Some of these data indicate that there are sources of orbital debris that are not understood. Consequently, more data are required.

Hazard assessment combines the environment with systems-damage criteria and hypervelocity-impact theory to determine the hazard to any particular spacecraft. A significant amount of hypervelocity testing was conducted during the Apollo program. This was testing of aluminum and glass structures at sizes and velocities that are directly applicable to orbital-debris issues. Currently, additional testing is being conducted into composite materials.

Space-object management consists of identifying and assessing the effectiveness of various control techniques, such as removal from orbit by retrieval or planned re-entry, placing a payload in a "disposal orbit," and minimizing the possibility of explosions in orbit. The last two techniques have already been used. That is, several payloads have been placed just outside geosynchronous orbit before the end of operational life. Also, new operational procedures have minimized the possibility of new Delta second-stage-rocket explosions. Other control techniques may prove to be expensive; consequently, it is important to understand the effectiveness of them before they become required policy. In view of what is known today, the following policies would be prudent:

1) Satellites should remain "intact" until re-entry. This would require that both payloads and rockets be designed to minimize the possibility of on-orbit explosions and that they retain structural integrity and intact thermal surfaces.

2) Population growth should be monitored of debris sizes smaller than those currently detected from the ground. This would not only provide a check for model predictions but would also detect any unmodeled sources and provide sufficient data either to eliminate the sources or to design shields to protect spacecraft from the measured environment.

3) Rocket stages used to transfer payloads from low Earth orbit to geosynchronous orbit should be re-entered within a year. When planned, the stages could easily be re-entered. Otherwise, they could remain in orbit for a very long time and pose a hazard to satellites in both low Earth orbit and geosynchronous orbit.

4) Any program that plans to place either large structures or a large number of objects in Earth orbit should be carefully examined for orbital-debris implications. Such programs, if not carefully conceived, could ini-

tiate a collisional fragmentation process that could cascade to other satellites.

Any other policy concerning orbital debris must await further evaluation. Although it is clear that certain future space activities may be jeopardized by orbital debris, it is not clear what limits are required to minimize the increase in debris while maintaining a growing space program. Policy to deal with these issues will require a complete understanding of debris, its propagation, and its effect on future space activities.

References

[1]Brooks, D. R., Gibson, G. G., and Bess, T. D., "Predicting the Probability that Earth-orbiting Spacecraft Will Collide with Man-made Objects in Space," *Space Safety and Rescue,* American Astronautical Society, Tarzana, CA, 1975, pp. 79–139.

[2]Kessler, D. J. and Cour-Palais, B. G., "Collision Frequency of Artificial Satellites: The Creation of a Debris Belt," *Journal of Geophysical Research,* Vol. 83, No. A6, 1978, pp. 2637–2646.

[3]Kessler, D. J., "Sources of Orbital Debris and the Projected Environment for Future Spacecraft," *Journal of Spacecraft and Rockets,* Vol. 18, July–Aug. 1981, pp. 357–360.

[4]AIAA Technical Committee on Space Systems, Space Debris, *An AIAA Position Paper,* American Institute of Aeronautics and Astronautics, New York, 1981.

[5]Kessler, D. J., "Summary of Orbital Debris Workshop," *Space Safety and Rescue 1982-1983,* G. W. Heath (ed.) American Astronautical Society, San Diego, CA, 1984, pp. 3–10.

[6]Kessler, D. J., "Orbital Debris Issues," *Space Debris, Asteroids and Satellite Orbits,* D. J. Kessler, E. Grun, and L. Sehnal, (eds.), *Advances in Space Research,* Vol. 5, No. 2, Pergamon, Oxford, England, 1985, pp. 3–10.

[7]NASA Office of Public Affairs, Satellite Situation Report, Vol. 27, No. 4, NASA Goddard Space Flight Center, Greenbelt, MD, 1987.

[8]Johnson, N. L., "History and Consequences of On-orbit Breakups," *Space Debris, Asteroids and Satellite Orbits,* D. J. Kessler, E. Grun, and L. Sehnal (eds.), *Advances in Space Research,* Vol. 5, No. 2, Pergamon, Oxford, England, 1985, pp. 11–19.

[9]Mueller, A. C. and Kessler, D. J., "Effects of Particulates from Solid Rocket Motors Fired in Space," *Space Debris, Asteroids and Satellite Orbits,* D. J. Kessler, E. Crun, and L. Sehnal (eds.), *Advances in Space Research,* Vol. 5, No. 2, Pergamon, Oxford, England, 1985, pp. 77–86.

[10]Su, S. Y. and Kessler, D. J., "Contribution of Explosion and Future Collision Fragments to the Orbital Debris Environment," *Space Debris, Asteroids and Satellite Orbits,* D. J. Kessler, E. Grun, and L. Sehnal (eds.), *Advances in Space Research,* Vol. 5, No. 2, Pergamon, Oxford, England, 1985, pp. 25–34.

[11]Hallgren, D. S. and Hemenway, C. L., "Analysis of Impact Craters from the S-149 Skyland Experiment," *Lecture Notes in Physics,* Springer-Verlag, Berlin, Heidelberg, New York, 1976, pp. 270–274; also K. Nagel, H. Fechtig, E. Schneider, G. Neukum, "Micrometeorite Impact Craters on Skylab Experiment S-149," *Lecture Notes in Physics,* Springer-Verlag, Berlin, Heidelberg, New York, 1976, pp. 275–278.

[12]Clanton, U. S., Zook, H. A., and Schultz, R. A., "Hypervelocity Impacts on Skylab IV/Apollo Windows," *Proceedings of the 11th Lunar and Planetary Science Conference,* 1980, pp. 2261–2273.

[13]Kessler, D. J., "Explorer 46 Meteoroid Bumper Experiment: Earth Orbital Debris Interpretation," presented at the IAU Colloquium on Properties and Interactions of Interplanetary Dust, Marseilles, France, July 9-12, 1984.

[14]Kessler, D. J., Zook, H. A., Potter, A. E., McKay, D. S., Clanton, U. S., Warren, J. L., Watts, L. A., Schultz, R. A., Schramm, L. S., Wentworth, S. J., and Robinson, G. A., "Examination of Returned Solar-Max Surfaces for Impacting Orbital Debris and Meteoroids," *Proceedings of the 16th Lunar and Planetary Science Conference,* 1985, pp. 434–435.

[15]Taff, L. G., Betty, P. E., Yakutis, A. J., and Randall, P. M. S., "Low Altitude, One Centimeter, Space Debris Search at Lincoln Laboratory's (M. I. T.) Experimental Test System," *Space Debris, Asteroids and Satellite Orbits,* D. J. Kessler, E. Grun, and L. Sehnal (eds.), *Advances in Space Research,* Vol. 5, No. 2, Pergamon, Oxford, England, 1985, pp. 35–45.

[16]Zook, H. A., Flaherty, R. E., and Kessler, D. J., "Meteoroid Impacts on the Gemini Windows," *Planetary and Space Sciences,* Vol. 18, Pergamon, Oxford, England, 1970, pp. 953–964.

[17]Kuczera, H., Iglseder, H., Weishaupt, V., and Igenbergs, E., "Acoustic Penetration and Impact Detector," *Space Debris, Asteroids and Satellite Orbits,* D. J. Kessler, E. Grun, and L. Sehnal (eds.), *Advances in Space Research,* Vol. 5, No. 2, Pergamon, Oxford, England, 1985, pp. 91–94.

[18]Sanguinet, J. A., "Satellite-Based Instrument Concepts for the Measurement of Orbital Debris," *Space Debris, Asteroids and Satellite Orbits,* D. J. Kessler, E. Grun, and L. Sehnal (eds.), *Advances in Space Research,* Vol. 5, No. 2, Pergamon, Oxford, England, 1985, pp. 59–62.

[19]Reynolds, R. C., "Design Considerations for Orbiting Detectors for the Large Low-Earth Orbit Debris Population," *Space Debris, Asteroids and Satellite Orbits,* D. J. Kessler, E. Grun, and L. Sehnal (eds.), *Advances in Space Research,* Vol. 5, No. 2, Pergamon, Oxford, England, 1985, pp. 63–75.

Chapter 2

Evolution of the Artificial Earth Satellite Environment

Nicholas L. Johnson*
Teledyne Brown Engineering, Colorado Springs, Colorado

Following the launch of Sputnik 1 in October 1957, the United States and the Soviet Union undertook vigorous space developments to assure reliable access to space and ability to deploy scientific and applications spacecraft. After the first decade of the space age, the international annual space launch rate had reached 127 missions. Contrary to popular belief, however, during the next two decades the annual space launch rate has remained virtually unchanged (maximum number of space launches in any one year, 129 during 1984).

Two common gauges of the growth of the Earth-satellite population are the size of the NORAD Satellite Catalog and the number of known satellites in Earth orbits. The NORAD Satellite Catalog compiles *every* identified Earth satellite since Sputnik 1. Under the United Nations Convention on Registration of Objects Launched into Outer Space, NORAD assigns identification numbers to each known space object. By January 1, 1987, this catalog had grown to 17,263 entries. During the past dozen years the catalog has expanded at about 780 objects per year.

Unfortunately, the NORAD Satellite Catalog does not explicitly provide a sense of the satellite population at a given time, because satellites are not purged from the catalog upon leaving Earth orbit, either by decay or injection into an Earth-escape trajectory. To remedy this situation, Teledyne Brown Engineering calculated the actual number of known satellites in orbit about the Earth at the end of each calendar year (Fig. 1). At the start of 1987, 6084 cataloged satellites were still circling the Earth.

Figure 1 clearly identifies two important factors affecting the future of the Earth-satellite population. The first is the noticeable decrease in the satellite population in 1979 and 1980. Although 900 new satellites were cataloged during this period, the true population declined by 300 satellites, owing to effects of the solar sunspot maximum (increase in atmospheric density at higher altitudes, consequently higher decay rates). Solar maxi-

*Advisory Scientist

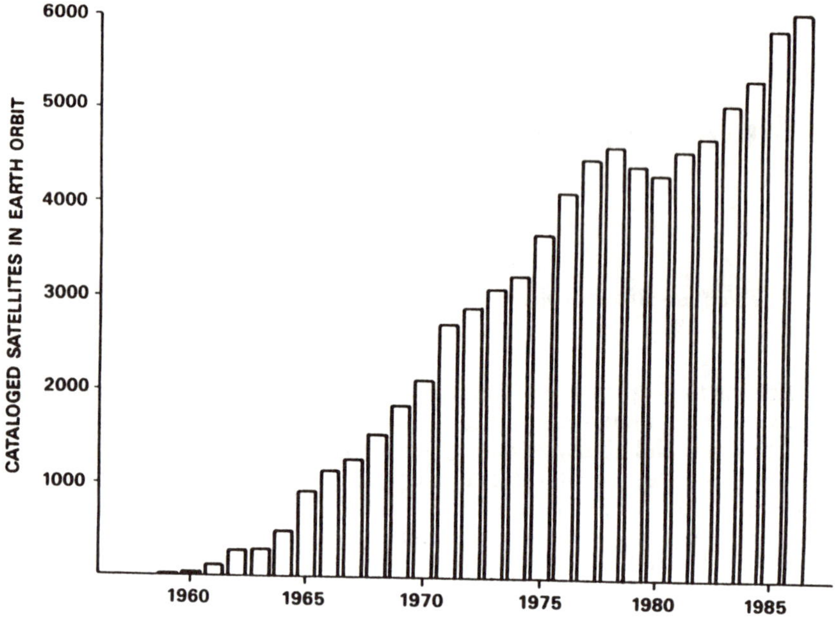

Fig. 1 Cataloged satellite growth.

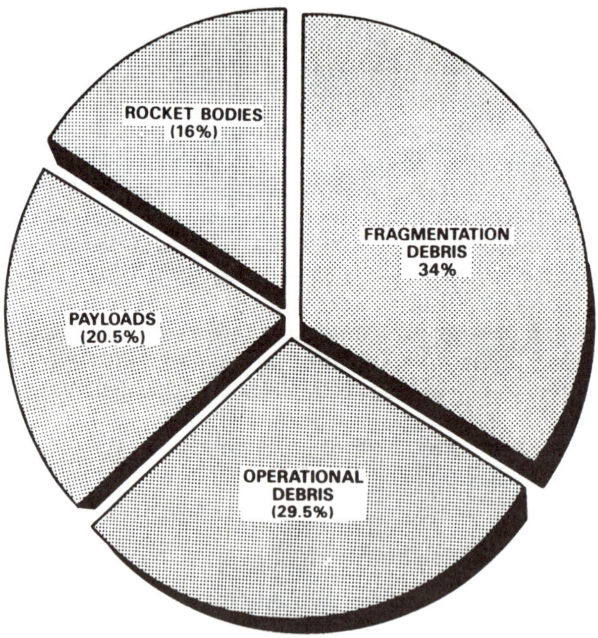

Fig. 2 Satellites cataloged: October 4, 1957–January 1, 1987.

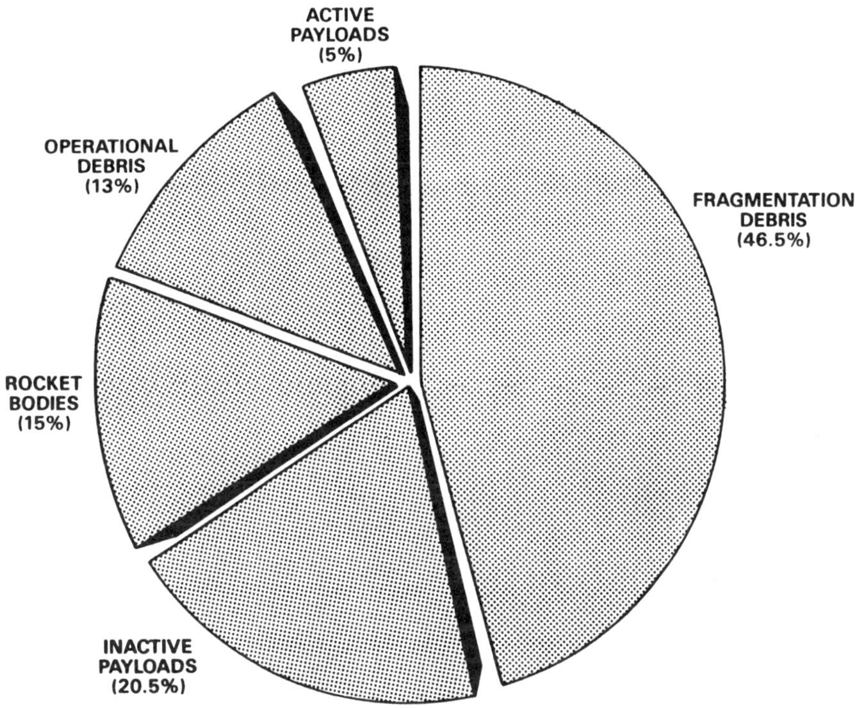

Fig. 3 Source of January 1, 1987, cataloged debris.

mums should be viewed as a periodically occurring, natural sink which will aid in cleansing the near-Earth environment.

The second factor is the actual rate of increase of the known satellite population. In the five years before solar maximum (1974—1978), the satellite population increased about 300 objects per year. In the six years following solar maximum (1981—1986), the growth rate was virtually unchanged. However, recent changes in spacecraft deployment philosophy may signal an increase in the growth rate.

In 1975, 21% of all Earth-orbit missions deposited satellites into "deep space" orbits—with periods greater than 225 minutes. By 1985, satellites in this category had jumped to 31%. Since satellites in high-period orbits are less likely to decay quickly, the satellite population may experience a higher growth rate during the latter half of the 1980's.

Another factor affecting the known Earth-satellite population is the recent improvement in the sensitivity of the NORAD space surveillance network. Objects which were heretofore undetectable by the network are now being tracked and cataloged. This greater sensitivity was partially responsible for the near-record 529 satellite growth registered in 1985.

Satellite decays can be caused by natural forces (e.g., atmospheric drag and solar-lunar perturbations) or by deliberate action as in the recovery of manned spacecraft. During the 1981–1985 period, the average number of

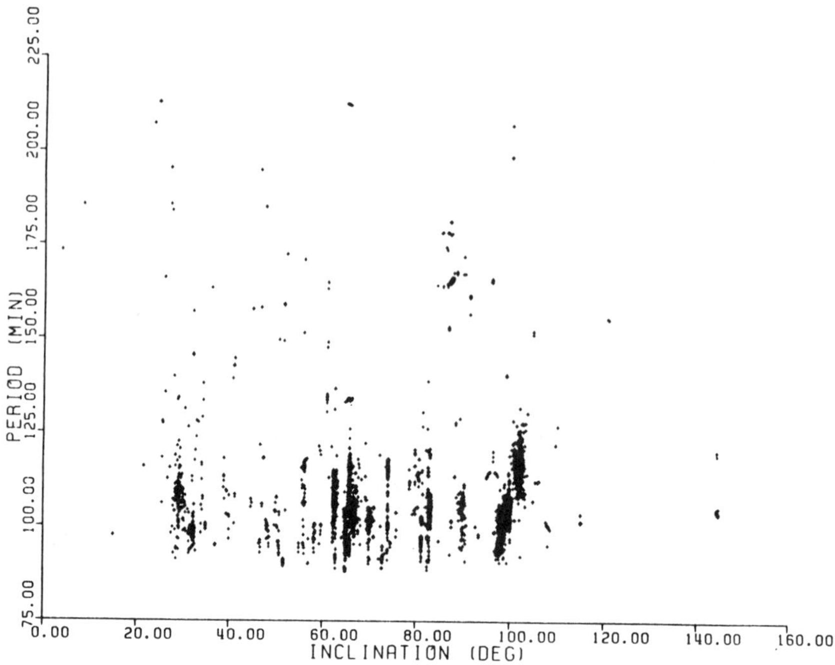

Fig. 4 Near-Earth satellite distribution.

cataloged decays per year was 548, i.e., three every two days. The recent decrease in the rate of decay probably reflects both the onset of solar minimum and the greater use of high-altitude orbits.

Status of the Current Satellite Population

As noted, on January 1, 1987 the cataloged satellites in orbit about the Earth totaled 6084. The true known population was actually greater because many other satellites were being tracked by NORAD but had not yet been entered into the official NORAD Satellite Catalog. A large number of satellites routinely remain in this limbo while NORAD confirms its ability to confidently track each object and establish its launch origin. Almost without exception these satellites represent pieces from satellite fragmentations or normal operational debris.

Since Sputnik 1, all cataloged satellites can be categorized into four general classes: payloads, rocket bodies, operational debris, and fragmentation debris. Operational debris includes objects released either intentionally (lens covers, ejection springs) or accidentally (gloves, screwdrivers) during normal mission operations. Figure 2 shows the relative contributions by each class. The current population is characterized by Fig. 3, which differentiates between active and inactive payloads. One can readily see that active payloads make up a very small percentage of the cataloged objects now circling the Earth.

The vast majority of Earth satellites, about 83%, reside in orbits with periods less than 225 minutes. Figure 4 gives some idea of the distribution. Some of the streaks evidence normal decay for specific, heavily-used missions; but most have arisen from satellite fragmentations, which litter a narrow-inclination band at widely ranging altitudes. The potential hazard of this environment to both manned and unmanned spacecraft operations can be quantified by examining the present and probable growth of the spatial density as a function of altitude. Spatial densities have been calculated using complete NORAD satellite databases for each 50-km-altitude band from 100 to 36,000 km at yearly intervals for the period 1977–87.

Effects of Satellite Fragmentations

The importance of the contribution of satellite fragmentations—either by accident or design—to the near-Earth environment cannot be over emphasized. On January 1, 1987, 46.5% of all known Earth satellites still in orbit had originated from satellite fragmentations. During the 90 + recognized satellite fragmentations since 1961, almost 5900 trackable pieces of debris were created, representing more than one-third of all satellites cataloged since Sputnik 1 (Fig. 5).

Of particular concern, such fragmentations are in good part preventable. The causes of satellite fragmentations can loosely be divided into three categories: propulsion-related accidents, deliberate detonations, and unknown causes (Fig. 6). Figure 7 breaks out the causes of known satellite fragmentation by five-year intervals. Immediately apparent is the significant increase of satellite fragmentations during the past decade. The largest factor in this increase is deliberate, and hence avoidable, satellite fragmen-

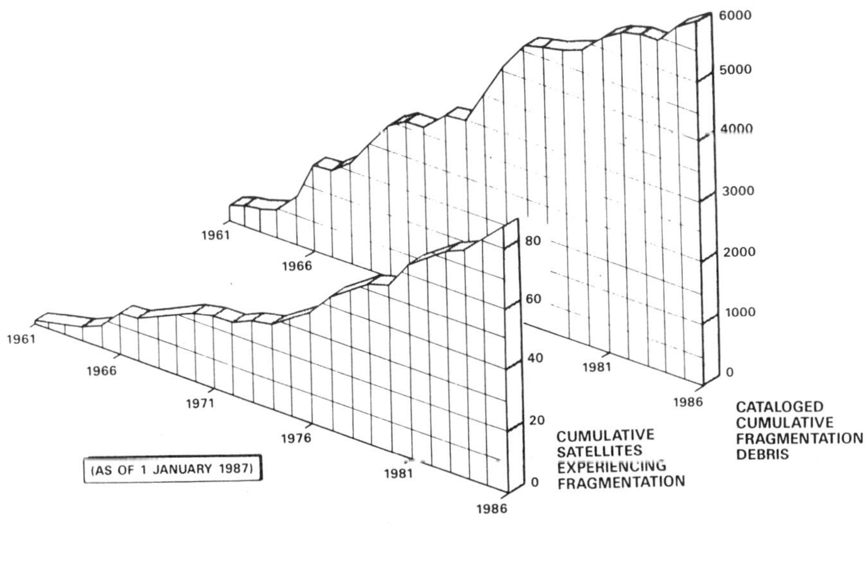

Fig. 5 History of on-orbit fragmentations.

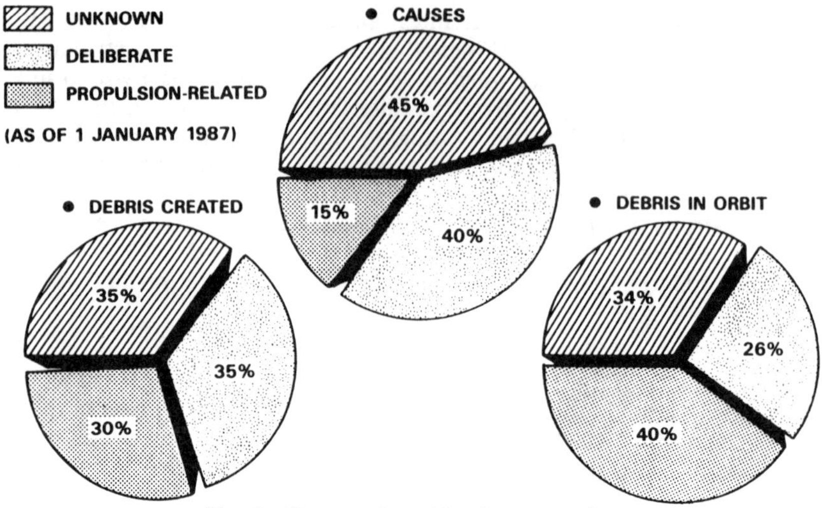

Fig. 6 Causes of satellite fragmentations.

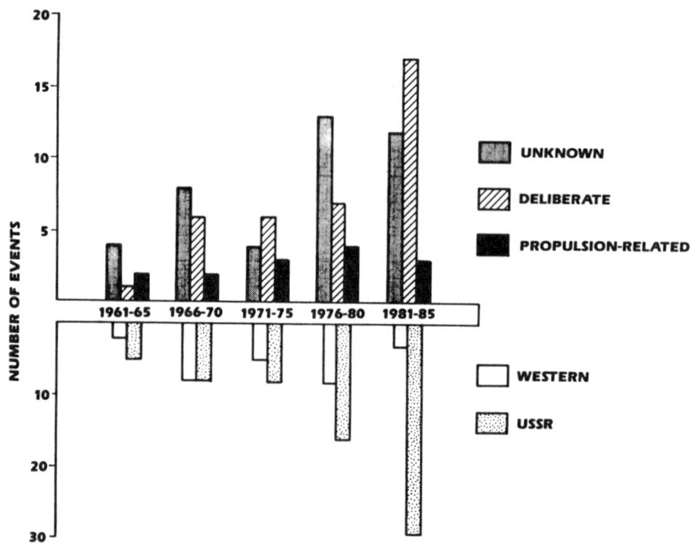

Fig. 7 History of satellite breakup events.

tations. Propulsion-related breakups occur on the average of once every two years and on the average contribute more debris per fragmentation than payloads (Fig. 8). Occasional events of this type are probably unavoidable.

The most densely populated altitude regime in 1977 was between 800 and 850 km with a spatial density of 1.49×10^{-8} satellites/km^3. By 1987 the region of greatest spatial density had shifted to 950-1000-km altitude, primarily as the result of a single satellite fragmentation, the USSR Kosmos

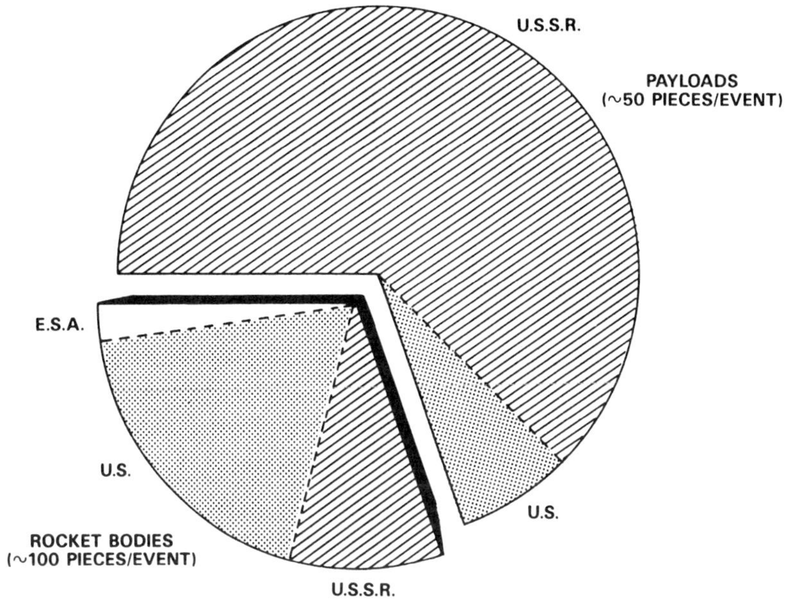

Fig. 8 Satellite fragmentations.

1275 (1981-53A, 12504). Approximately one-third of all known satellites now penetrate this region. Whereas the spatial density in the 800-850-km-altitude regime increased by 41% between 1977 and 1987, the spatial density in the 950-1000-km-altitude regime increased 128% to 2.73×10^{-8} satellites/km^3. If the fragmentation of Kosmos 1275 was the result of a hypervelocity collision with an untracked object, as some evidence suggests, the number of new particles with a diameter of 1 mm or greater would raise the spatial density by several orders of magnitude.

The expanded use of deep-space orbits, particularly semi-synchronous and geosynchronous orbits, measureably raise probability of collision at higher altitudes. Since 1977 the spatial density at geosynchronous altitude has increased by more than two orders of magnitude (from 1.85×10^{-11} to 2.38×10^{-9}), although this region still remains less hazardous than low Earth orbits. Likewise, the spatial density in the altitude regime 20,100 to 20,150 km increased an order of magnitude (9.23×10^{-13} to 9.52×10^{-12}), primarily due to the initiation of the U.S. NAVSTAR Global Positioning System (GPS). The other major contribution at high altitudes has been the accumulation the bodies of geosynchronous transfer rockets.

Breakup debris now resides in a small number of selected regions defined by altitude and inclination (Fig. 9). Unfortunately, the orbital planes of debris are for the most part randomly distributed about the Earth. Therefore, the debris actually poses a hazard to virtually *all* other satellites, operational and nonoperational (Fig. 10). The ultimate hazard of satellite-breakup debris is collision with larger space objects and the subsequent creation of great numbers of secondary debris.

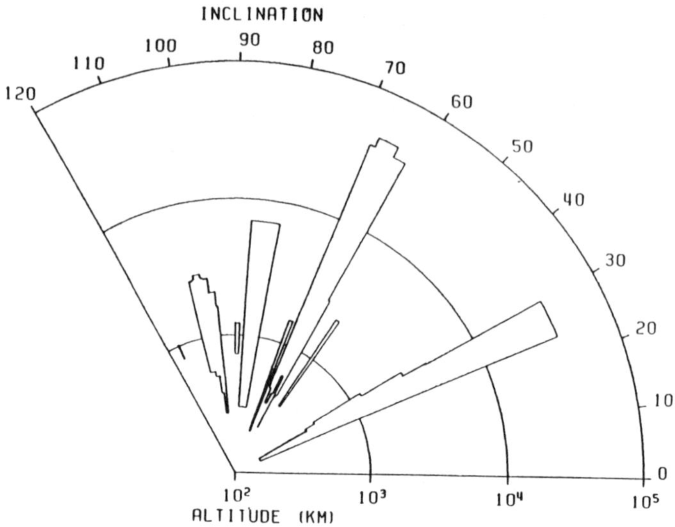

Fig. 9 Distribution of fragmentation debris.

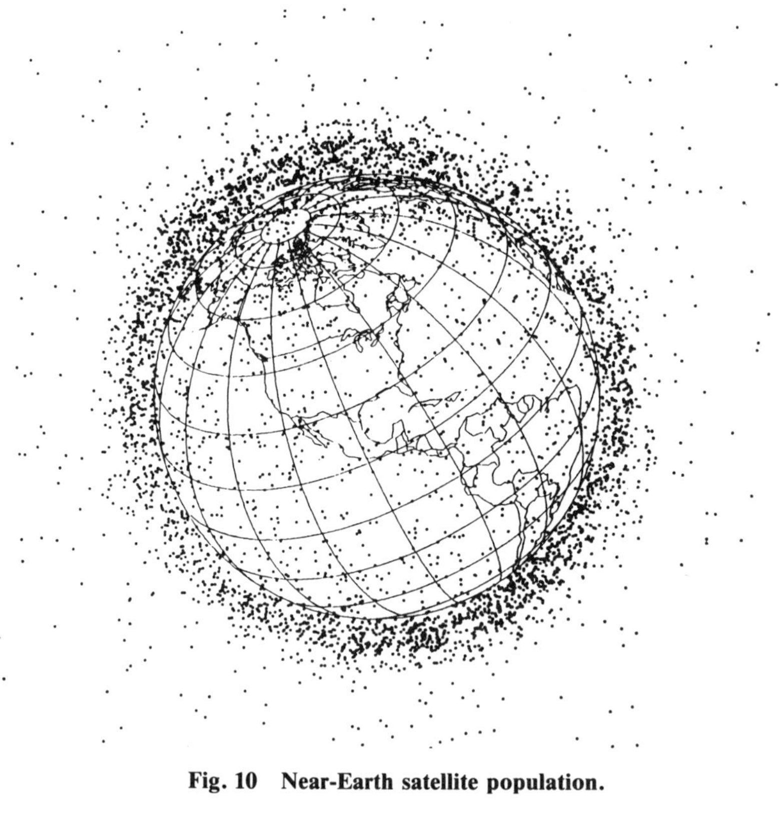

Fig. 10 Near-Earth satellite population.

On November 13, 1986, the SPOT/Viking rocket body (1986-19C, 16615) fragmented into almost 500 detectable pieces, of which more than 330 had been cataloged by July 1, 1987. From the original orbit of 804 by 833 km at an inclination of 98.7°, the vast majority of the resultant debris was thrown into very-long-lived orbits. Some of this debris already traverses the altitudes frequented by manned spacecraft, and the rest will eventually decay through this regime. While the cause of the fragmentation is still under investigation, a re-examination of previous Ariane third-stage fragmentations is being conducted. The low-inclination, high-period orbits of the earlier events have hindered the assessment of a causative agent; but identifying it will be important both to inform corrective action and to evaluate better the probable numbers of unseen debris and the consequent effect on the local spatial densities.

Chapter 3

Hypervelocity Impact and Upper-Stage Breakups

B. G. Cour-Palais* and J. L. Crews†
NASA Johnson Space Center, Houston, Texas

Some limited tests have been made to determine whether a distinct fragmentation signature will allow discrimination between hypervelocity impacts and internal explosions for launch-vehicle upper stages. The tests and the subsequent analyses were performed on very short notice. This led to the rather unusual procedure of using aluminum soft drink cans to represent an upper-stage tank. It will be shown that the cans used were a reasonable representation of a typical launch-vehicle upper stage. We discuss the pertinent factors controlling the damage patterns caused by hypervelocity impacts, compare results for pressurized and unpressurized cans, and make observations as to the applicability of this work to actual upper stages.

Controlling Factors in Hypervelocity Impacts

The upper-stage tank for a liquid propellant launch vehicle—for the most part, a thin-walled, large-diameter cylindrical vessel—has two compartments separated by a bulkhead, i.e., the fuel and oxidizer tanks. The ends of the cylinder are closed off by hemispherical bulkheads, with the engine assembly on one end and the guidance and telemetry "black boxes" on the other. One such upper stage, the third stage (Fig. 1) of the Ariane 2 and 3 launch vehicles, has a diameter of 2.6 m and a length of 8.12 m between the apex of each bulkhead. The average thickness of the cylindrical walls has been calculated to be 0.13 cm (0.052 in.) from the skin pressure-stress levels quoted by Arianespace engineers.[1] In effect, an empty cylindrical upper stage functions as a "bumper" when hit by a hypervelocity orbital-debris fragment. This means that the first surface hit shatters the fragment into an expanding cloud of solid or liquid particles that moves through the tank until it meets the other wall of the cylinder (see Fig. 2). Its effect on the other side of the tank depends on a number of factors.

*Senior Scientist, Space Physics Branch.

†Manager, Hypervelocity Impact Research Laboratory.

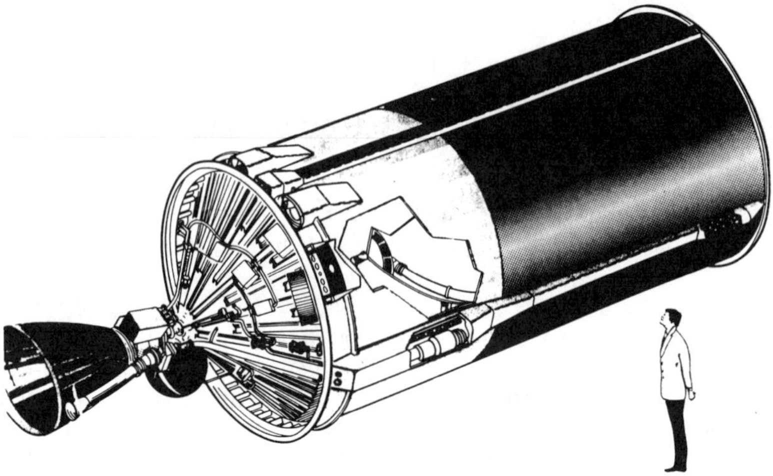

Fig. 1 Ariane 2 and 3 third stage.

The degree of fragmentation of an impacting bit of debris and its physical state depends on the ratio of the mass per unit area of the tank wall to that of the debris. If both the debris and the wall have the same material properties, as they do in this case, the controlling parameter becomes the ratio of the tank thickness t_w to the debris diameter d_p. Another controlling factor is the relative impact velocity V_p of the incoming bit of debris.

Spacing in a double-wall container also has a strong effect on the results of an impact. If the projectile and shield are in a liquid or vaporized state, the expanding cloud imparts an impulsive load to the second wall over an area that depends on the spacing and the impact velocity. In addition, the impulse decreases as the inverse square of the spacing. On the other hand, if the projectile and/or the shield plug remain solid fragments after the primary impact, the effect of spacing will be less pronounced. If the medium is the very low vacuum of space, a fragment will not be decelerated and will punch a hole in or crater the second wall, which may lead to a large hole. These two types of damage and the equations used to determine the thickness of the second sheet are discussed in Ref. 2.

The remaining factors controlling the two-wall impact results are the material properties involved in the event: mass density, ductility, compressive and tensile yield-strengths, modulus of elasticity, melting and vaporization temperatures, and latent heats. The pressures generated at impact are a function of the mass densities of the two bodies in collision and the relative impact velocity. The impact pressure for aluminum to completely melt is 0.9 megabars; this can occur when two aluminum surfaces collide at 5.6 km/s.[3] The onset of aluminum vaporization will occur at 1.67 megabars at an impact velocity of 10.2 km/s.[3] The thermal conditioning of the projectile and target is caused by the trapped entropy arising out of rapid compression and expansion during impact.

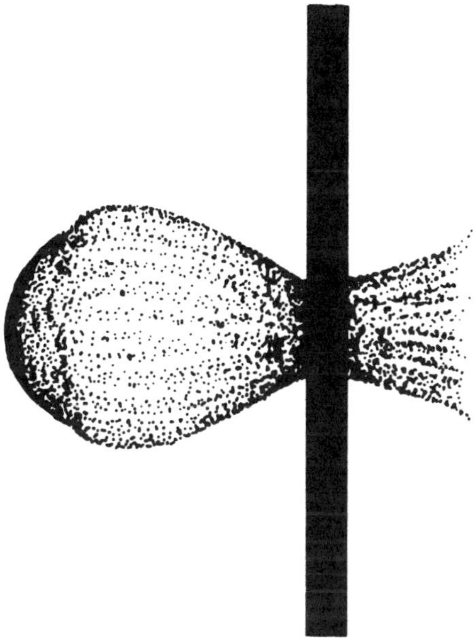

Fig. 2 Illustration of debris formed when a projectile impacts a thin sheet at hypervelocity.

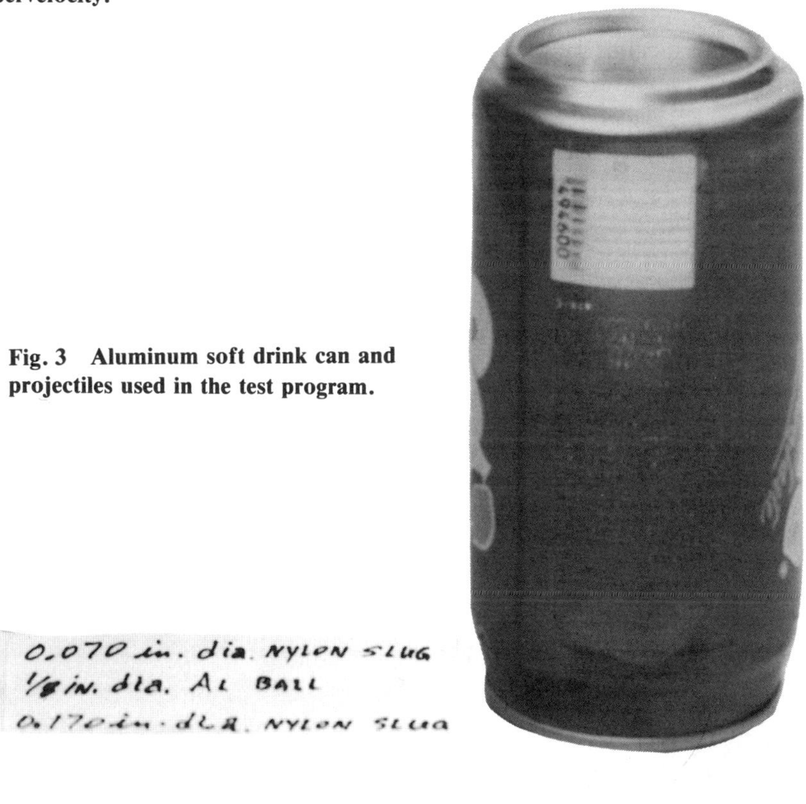

Fig. 3 Aluminum soft drink can and projectiles used in the test program.

Table 1 Can hypervelocity-impact data

Lab#	Projectile density, R_p, gm/cc	Projectile diam, D_p, cm	Projectile length, L_p, cm	Projectile mass, M_p, gm	Projectile velocity, V_p, km/s	Target mass initial M_i, gm	M_p/M_i, %	Target mass loss, gm	Major fragments	Major fragments mass, gm
1112	2.780	0.178	0.081	0.005	6.57	19.04	0.026	0.00	1	19.04
A186	1.200	0.428	0.436	0.070	6.95	16.59	0.420	0.48	1	15.99
1114	2.780	0.180	0.079	0.005	6.50	22.35	0.022	0.00	2	22.28
1115	2.780	0.177	0.084	0.005	6.95	78.35	0.006	1.95	1	76.40
1116	2.780	0.177	0.086	0.005	6.95	18.28	0.027	0.00	3	17.72
A191	2.780	0.318	0.318	0.043	6.74	18.09	0.240	0.00	3	17.45
A192	2.780	0.318	0.318	0.045	6.48	25.11	0.180	0.00	3	24.72

Lab#	Debris	Debris mass, gm	Dust mass, gm	Comments
1112	0	0.00	0.00	Entry hole: .2 cm × .15 cm; exit hole: 1 cm × 1.2 cm; petalled
A186	8	0.04	0.02	Entry hole: .65 cm × .45 cm; exit hole: 5.5 cm × 4.5 cm; witness plate: 2 deep craters
1114	0	0.00	0.07	Press: 1 ATM; target torn open and sheared into two pieces; some small debris
1115	0	0.00	0.00	Press: 1 ATM + water; no rupture, bulge; particles in H_2O; mass loss is H_2O
1116	0	0.00	0.08	Press: 1 ATM; target torn open and sheared into two pieces; some small debris
A191	19	0.12	0.10	Press: 1 ATM; target torn open and sheared into three pieces; some small debris
A192	11	0.13	0.10	Press: 1 ATM + plastic liner; target in 3 pieces; sooty deposit; small debris

Upper-Stag Scale Model Tests

In the time allowed for this study, it was not possible to have aluminum cylinders made to represent an upper-stage tank exactly to scale. However, it turned out that the typical aluminum soft drink can (Fig. 3) offered a reasonable representation of upper stages. The can used had an external diameter of 7.22 cm, an overall length of 12.38 cm, and a 0.013-cm skin thickness over the cylindrical section. It was drawn to a thickness of 0.032 in. from 3003-H19 aluminum sheet, the base integral with the body. The "pop-top" portion was crimped over the body of the can sometime after the drawing process. The cans used in these tests weighed between 16.6 and 19 g empty, and when resealed could hold an internal pressure differential of 14.7 psi without leaking.

The ratio of the radius R to skin thickness t_w of a cylinder conditions the effect of curvature on the impact results relative to that of a flat plate. As most of the experience with the behavior of two-sheet structures in hypervelocity impact testing has been with flat targets, it is necessary to know whether the cylindrical case being investigated is directly comparable or not. The value of R/t_w for the cans is 284 and for the Ariane upper stage of Fig. 1, 277. Both numbers are large enough to make minimal any radius effects on the impact results, and the can data can thus be taken as representative of the full-size tank.

The aspect ratio (overall length-to-diameter ratio of the can) is 1.72, which is not very different from the hydrogen tank of the Ariane upper stage between the bulkheads, which is about 2.0. This parameter establishes the geometrical similarity between the test and the real upper stage, and does not influence the impact results. For example, if the can had an aspect ratio very much less than that for the upper stage, the expanding debris vapor could have been affected by reflection from the ends. Finally, the skin thickness of the can is one-tenth that of the calculated average Ariane skin thickness.

Seven tests were conducted. These involved two unpressurized cans, three cans pressurized to 1 atm with air, one can 1/5th fllled with water and pressurized to 1 atm, and one can internally lined with a thin layer of foam and also pressurized.

The tests were performed in the NASA Johnson's Hypervelocity Impact Research Laboratory using two different light-gas guns. One of these guns, with a 0.17-cm bore, was used to fire 5-mg aluminum projectiles at speeds between 6.5 and 7 km/s. The other light-gas gun, with a bore of 0.43 cm, can launch 70-mg solid plastic projectiles in this same velocity range. For these tests, this second gun was also used to fire 0.32-cm, 45-mg spherical aluminum projectiles using the sabot technique.

The gun ranges are totally enclosed and were evacuated to 1/1000 atm for the tests. The aluminum cans were suspended horizontally in the impact chamber by fine thread and arranged so that projectiles hit close to the perpendicular. The velocity of the projectile just prior to the impact was obtained using a rotating-mirror camera that records 80 frames of a transient event at 10^6 fr/s. A video camera was used to record the impact on the cans for later playback. Table 1 gives test data.

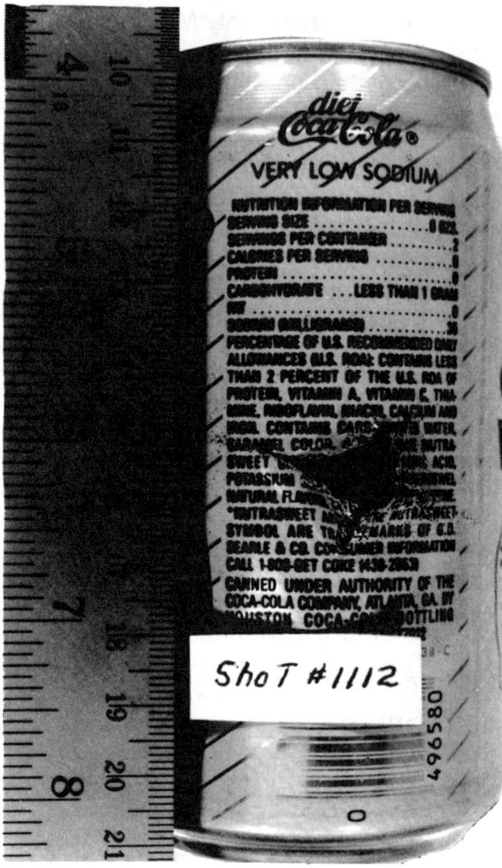

Fig. 4 Can impact damage from a 5-mg aluminum projectile at 6.57 km/s.

Test Results and Analysis

The results of some of the tests can be seen in Figs. 4–8, and may be summarized as follows:

1) The unpressurized cans remained essentially intact, with a petalled hole blown open on the side opposite the entry point of the projectile. Figure 4 shows this for a 5-mg projectile; Fig. 5 for the 70-mg projectile. The fragments collected after the latter impact are shown in Fig. 6; they weighed 0.48 g. A witness plate behind the can showed that part of the 70-mg projectile survived the encounter with the can.

2) The pressurized cans were catastrophically ruptured into two or three major pieces (Fig. 7, Table 2). Only very minor fragments were gathered up after these shots.

3) The pressurized can partially filled with water remained intact, with only a bulge opposite the projectile's entry point. The momentum of the projectile was completely absorbed and all of the impact debris was con-

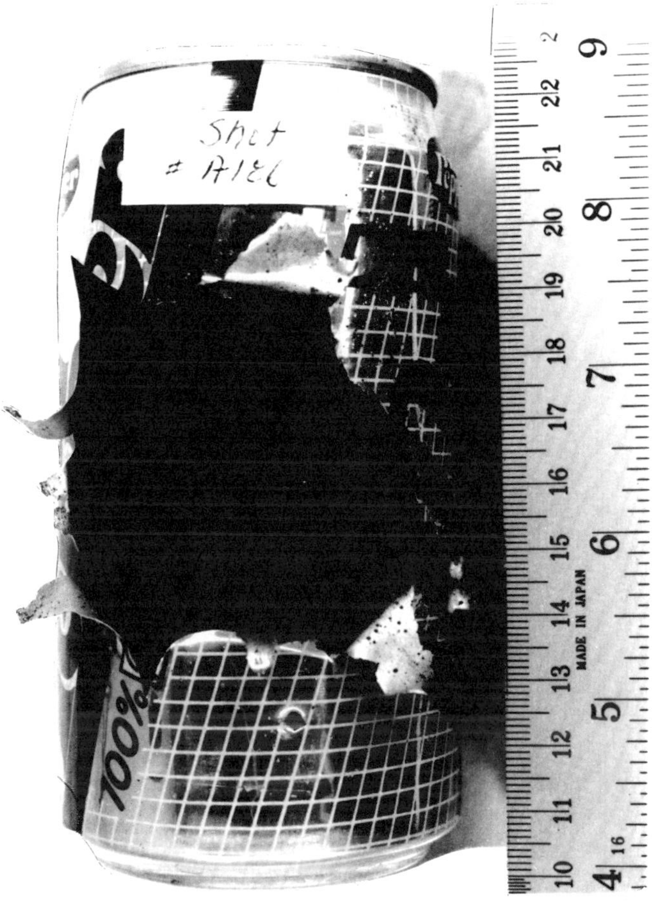

Fig. 5 Can impact damage resulting from a 70-mg aluminum projectile at 6.95 km/s.

tained by the water. The motion of the can was detected by the video camera, and its velocity was calculated to be about 68 cm/s. Another interesting development was that the water inside the can spurted out of the entry hole for several seconds as the pressures were equalizing.

4) The pressurized can with the liner ruptured like the others and showed evidence of darkening of the bright aluminum surfaces due to localized burning of the foam. This can be seen in Fig. 8 as a darkening of the aluminum surface in the bottom piece of the can.

One phenomenon that deserves mentioning was the very bright impact flash that was detected by the video camera on several of the shots. The existence of an impact flash has been known for a long time. The open literature describes results of several flash studies.[4]

Table 2 Can fragment distributions

Test # M_i	Projectile mass, M_p, gm	Initial target mass, M_i, gm	M_p/M_i, %	Major fragments mass, M_a, gm	Major fragments number	M_a/M_i	Medium fragments mass, M_p, gm	Medium fragments number	M_b/M_i	Minor fragments mass, M_c, gm	Minor fragments number	$M_b/$
A186	0.070	16.59	0.422	15.99	1	0.964	0.42	4	0.025	0.04	4	0.002
1114	0.005	22.35	0.022	22.28	2	0.997	0.00	0	0.000	0.00	0	0.000
1116	0.005	18.28	0.027	17.72	3	0.969	0.00	0	0.000	0.00	0	0.000
A191	0.043	18.09	0.238	17.45	3	0.965	0.07	6	0.004	0.04	13	0.002
A192	0.045	25.11	0.179	24.72	3	0.984	0.13	1	0.005	0.04	10	0.002

Test #	Dust Fragments mass, M_d; gm	M_d/M_i
A186	0.02	0.001
1114	0.07	0.003
1116	0.08	0.004
A191	0.1	0.006
A192	0.1	0.004

Fig. 6 Fragments from can hit by the 70-mg projectile.

Application of Results

Application of data obtained in the tests to a third-stage booster tank such as the Ariane is possible both qualitatively and quantitatively. The very large scale-factor complicates quantitative interpretation.

Two scaling relationships will be assumed to be constant over the size range involved, the ratios M_p to M_i and t_w to d_p, where M_p and M_i are the initial masses of the projectile and can or tank stage, respectively, t_w the thickness of the walls, and d_p the diameter of the projectiles. For the unpressurized tests (#1112 and #A186 in Table 1), there are two values of M_p/M_t and that correspond to the different damage conditions. The minimal damage shown in Fig. 4, with the can ruptured but still in one piece, was obtained for a projectile mass 0.026% of the can mass, and a t_w/d_p ratio of 0.12. For the case shown in Fig. 5, where the can has been blasted open and sizable fragments ejected, M_p/M_t has a value of 0.42%, and t_w/d_p is equal to 0.03. Using a value of 1115 kg for the mass of the Ariane third-stage tank, i.e., total stage weight without the Vehicle Equipment Bay and the clamp ring, the debris masses that could cause similar damage to a tank like the Ariane's are approximately 0.3 and 4 kg, respectively, at a relative impact speed of 7 km/s.

The pressurized cans (one shown in Fig. 7) were typically torn apart into two or three large remnants. The maximum (hoop) stress in the cans at 1 atm was 4120 psi, which was about 25% of the tensile and 40% of the shear allowable stress for the aluminum can material. In each of the five tests, the

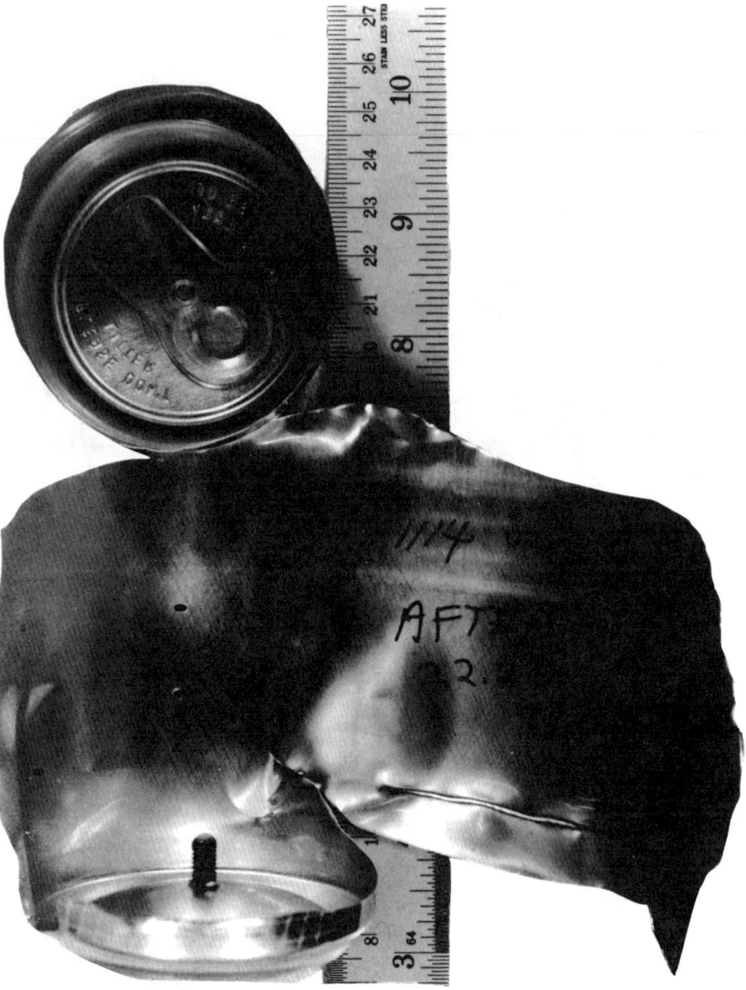

Fig. 7 Typical result for a pressurized can hit by 5- and 45-mg aluminum projectiles at speeds above 6.5 km/s.

projectile debris from the impact on the first surface initiated rupture of the opposite wall, like the one in Fig. 4, which caused it to tear open under the pressure strain energy. The minimum mass ratio M_p/M_t for the initiation of failure for the pressurized cans was 0.022%, corresponding to a debris mass equal to 0.25 kg to cause the same failure for the Ariane third stage. The maximum residual pressure stresses in the hydrogen and oxygen tank walls of this stage were given as 3000 bar(43,500 psi) and 2380 bar(34,500 psi), respectively, by the Arianespace engineers.[1] These values equate to approximately ten times the strain energy that was present in the cans, and the stresses are also much closer to the material allowables in the case of the

Fig. 8 Fragments from can with plastic internal liner showing blackening due to carbon deposit.

tank stage. Both these factors should cause a pressurized tank to fail by tearing rather than by fragmenting into small chunks.

Table 2 lists the fragment distributions associated with the larger of the two unpressurized tests and four of the pressurized tests. The fragment masses are expressed as proportions of the initial target masses, for four size ranges. Major fragments are the large remnants of the cans, like those seen in Figs. 4, 5, and 7, while medium-size fragments are as depicted in Fig. 6. Minor fragments are no larger than 5 mm^2, and the dust is about 1 mm^2 or smaller. It is obvious that in every case a few large pieces account for over 96% of the initial can masses. Only in the unpressurized case do we find a significant fraction of the initial mass in medium fragments. This is a result of the different failure mechanism involved. The minor fragments and the dust account for under 0.1% of the total mass.

The Ariane third-stage fragment masses were calculated from the radar cross sections, and the decelerations determined from at least three successive orbits for 424 fragments.[5] A portion of the data from the report is reproduced in Table 3. It is evident that the mass is concentrated in three very large and three considerably smaller fragments. The three big fragments account for 90.5% of the mass in the 424 fragments (1246.4 kg). The next 10 smaller fragments account for 6.0%. These 13 fragments together total 96.5%. Another 11 fragments totaling 1.40%, makes a total for the 24 fragments of 97.9%. The balance of the 400 fragments account for the remaining 2.1%.

Table 3 Ariane third-stage fragment distributions

Number	RCS, m^2	Mass, kg	Mass > 100 kg	Mass > 4 kg ≤ 100 kg	Mass > 1 kg ≤ 4 kg
16616	1.719	23.22		23.22	
17130	2.752	5.97		5.97	
17203	0.376	1.89			1.89
17206	1.422	2.78			2.78
17212	0.478	1.00			1.00
17594	13.640	5.15		5.15	
17116	0.968	1.07			1.07
17119	2.112	4.13		4.13	
17122	16.362	647.81	647.81		
17131	31.200	116.93	116.93		
17153	0.546	1.96			1.96
17201	0.316	1.30			1.30
17316	0.338	1.17			1.17
17324	2.293	4.50		4.50	
17123	1.050	6.94		6.94	
17129	11.422	363.69	363.69		
17154	0.274	1.12			1.12
17208	0.514	1.37			1.37
17223	5.152	9.12		9.12	
17333	1.330	5.35		5.35	
17425	1.195	6.25		6.25	
17457	7.083	4.16		4.16	
17472	3.700	1.39			1.39
17478	0.228	1.86			1.86
24		1220.13	1128.43	74.79	16.91

There is a difference of about 180 kg between the total calculated mass in the 424 fragments and the actual weight of the Ariane third stage in orbit before the breakup. It can be accounted for by errors (±30%) in the derivation of mass from radar cross sections.[5] For instance, fragment number 16616 in Table 3 refers to the Ariane clamp ring, its calculated mass 23.2 kg and the known mass 28 kg. However, the mass percentages quoted above should remain unchanged.

A comparison of the test results in Table 2 and the Ariane fragment distribution in Table 3 shows a marked similarity. In both cases, a few large fragments contain the major portion of the total mass. This signature would indicate that the Ariane breakup was triggered by an impact that initiated a tear or crack on the opposite side. The very high strain energy in the skin due to the residual pressure would have caused it to shear into large pieces, with very little fragmentation. An internally triggered explosion would have caused a more normally distributed fragment distribution, since there would be no single initiating hole or crack.

As for can tests qualitatively, it is obvious that even a moderate pressure stress level in a cylindrical vessel will cause catastrophic rupture, even when it is impacted at hypervelocity by a very small relative mass. Due to the nature of the failure mechanism, a pressurized can does not fragment into a sized distribution, but rather tears into a few large pieces. An unpressurized can, with zero wall stress, will fragment when hit by an object with a large enough fraction of the target mass. A partially filled and pressurized tank, on the basis of one test, absorbs the impact debris without rupture; but this condition probably has no parallel in the case of the Ariane stage tank.[6]

Conclusions

The conclusions that may be drawn from this series of tests pertain not only to the Ariane third stage but also to all tank stages remaining in Earth-orbit for a reasonably long time. If the stage is not vented after its mission, the residual pressure will cause strains that can lead to catastrophic destruction by an object of a relatively small mass hitting at speeds greater than 4 km/s. The number of pieces will not be many, because the stage will not burst, but rather tear along impact-related cracks (see Fig. 4, 5, and 7). When a tank is vented, an impact by an object of the same mass that causes a pressurized tank to rupture would not cause much more than a small opening on the opposite wall, with very little ejecta. Struck by an object a third the mass of the stage or larger, the tank would probably remain in one piece, but have a large hole on the side opposite the point of entry. Because a tank stage is essentially void of internal components, it will not fragment in the same manner as a satellite that is filled with electronic and other components. However, it is possible that the momentum exchange resulting from a large impacting mass may cause the externally attached engine and guidance components to break off and add to the number of fragments observed. Again, the addition of these pieces to the fragment distribution resulting from an impact will not change the overall conclusion: most of the total mass will be in a few fragments.

With specific reference to the breakup of the Ariane third-stage tank, the pressurized-can test data is consistent with the small number of large fragments observed using ground-based telescopes. Also, the fragment mass distribution matches.

Another question answered by this test series was whether plastic materials used in satellites impacted by a hypervelocity projectile would deposit on fragments. This had been postulated as a possible cause of failure to optically-observe satellite debris. The single test previously described confirmed that the usually bright aluminum fragment surfaces are indeed dulled after the impact. It is assumed that the high-temperature impact debris causes the plastics to burn and deposit carbon.

The scope of this test series was limited by the need for a quick response, but it provided enough information to judge whether or not a tank-stage breakup could be due to an impact. For the Ariane third stage, the possibility that the fragmentation which occurred was triggered by an impact event is more plausible than ever.

References

[1]Private communication with Arianespace engineers.

[2]Cour-Palais, B. G., Space Vehicle Meteoroid Shielding Design: ESA SP-153, Oct. 1979.

[3]Swift, H. F., *Impact Dynamics,* Zukas et al., p. 215., Wiley, New York, 1982.
[4]Jean, B. and Rollins, T. L., "Radiation from Hypervelocity Impact Generated Plasma," AIAA Paper 69-364, Hypervelocity Impact Conference, Cincinnati, OH, 1969.
[5]Badhwar, G. D. and Anz-Meador, P. D., "Determination of the Area and Mass Distribution of Orbital Debris Fragments," Presented at the International Astronautical Federation Meeting, Bangalore, India, Oct. 1988.
[6]Hergott, R. and Heloret, J. Y., "Ariane Third Stage: Current Operational Practice and Modifications Expected," published elsewhere in this volume.

Breakup Studies

Chapter 4

Preliminary Analysis of the Fragmentation of the Spot 1 Ariane Third Stage

Nicholas L. Johnson*
Teledyne Brown Engineering, Colorado Springs, Colorado

After circling the Earth in an apparently dormant state for 9 months, on November 13, 1986 the SPOT 1 Ariane third stage violently separated into some 465 detectable fragments—the most severe satellite breakup yet recorded. The vast majority of these debris will post a hazard to spacecraft operations for hundreds of years. Although not confidently identified, overpressurization of the rocket body from residual propellants might have caused the breakup. Evidence indicates that other Ariane third stages may also have brokenup, the effects being masked by difficult-to-observe orbits. The possibility that the SPOT 1 Ariane third stage was struck by an unseen object has not been ruled out. A curious asymmetry in the debris ejection pattern may help in establishing a probable cause.

This chapter explores the event.

SPOT 1 Mission Background

The sixteenth flight of the European Space Agency's Ariane launch vehicle was conducted on February 22, 1986. Unlike earlier missions, whose objectives were low-inclination (< 18°), highly-elliptical orbits, this one was designed to deposit two payloads into a high-inclination, Sunsynchronous orbit. The primary payload was the French SPOT 1, a 1750-kg Earth-resources satellite roughly analogous to the U.S. Landsat. Attached to SPOT 1 with a 28-kg clamp band was a piggyback payload, the 286-kg Swedish Viking satellite designed to study the ionosphere and magnetosphere with emphasis on auroral phenomena.

The Ariane 1 launch vehicle employed on this mission utilized a liquid oxygen/liquid hydrogen-powered third stage designated H8 (with reference to the 8 metric tons of propellants carried). The launch from Kourou, French Guiana, took place late on February 21, 1986 local time (early February 22, 1986 GMT). The payload was inserted into an orbit with a mean altitude of about 820 km at an inclination of 98.7° (Appendix 4).

---

*Advisory Scientist.

Table 1 Original SPOT 1 mission satellites

Satellite number	International designator	Identity
16613	1986-19A	SPOT 1
16614	1986-19B	Viking
16615	1986-19C	Ariance third stage
16616	1986-19D	Clamp band

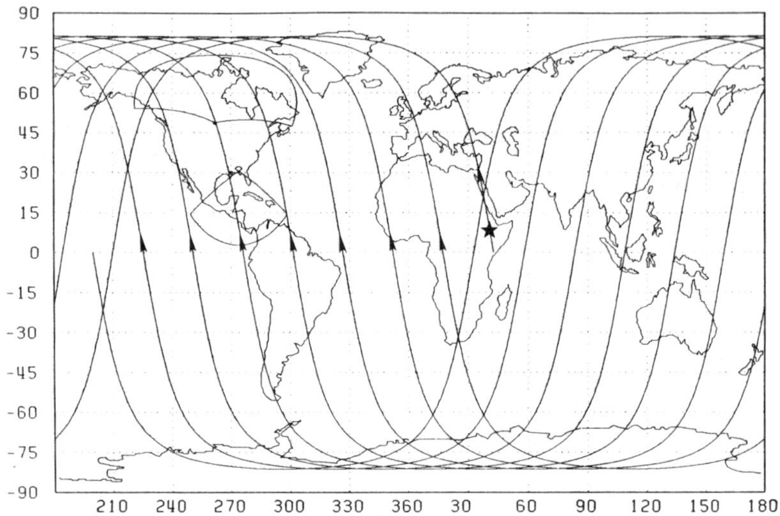

Fig. 1 Initial track (13½ h) of the SPOT 1 Ariane third-stage fragments. The debris cloud passed over the United States about 8 h after the fragmentation. The maps shows coverages (grey areas) of the PARCS and FPS-85 radars for 820-km altitude.

Shortly after shutdown of the HM7 engine of the Ariane third stage, three objects separated from the spent rocket body. Each object was assigned a satellite number and international designator by NORAD, as indicated in Table 1.

Using its on-board propulsion, the Viking satellite raised its apogee to over 13,500 km while retaining a perigee near 800 km. SPOT 1, the Ariane third stage, and the clamp band were left in a Sun-synchronous orbit with a time of descending node (southward crossing of the equator) at approximately 1030 local Sun time. The relevance of this nodal crossing time to the fragmentation will be discussed later. (Sun-synchronous orbits, having rate of nodal precision exactly matching the angular rate of the Earth about the Sun, cause a satellite to pass over a given latitude at the same local Sun time each day.)

The last official NORAD orbital element set for the Ariane third stage prior to November 13, 1986 was apparently issued on October 31, 1986, as

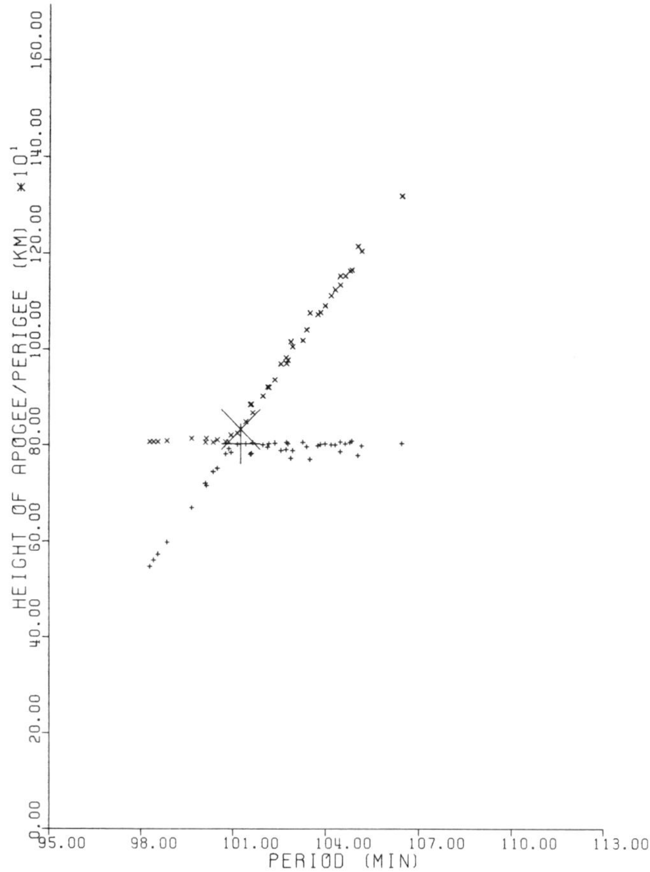

Fig. 2 The FPS-85 radar at Eglin AFB detected 44 fragments from the SPOT 1 Ariane third stage only 8½ h after the breakup.

summarized here, together with a radar cross-section (RCS) estimate from the FPS-85 radar (442 MHz) at Eglin Air Force Base (AFB) in Florida:

Epoch	1986, Day 304.66151240
Inclination, deg	98.6973
Eccentricity	0.0021179
Mean motion	14.22163579
Argument of perigee, deg	61.2744
Mean anomaly, deg	299.0538
Right ascension, deg	17.5922
Radar cross-section, m^2	27.54

Fragmentation of SPOT 1 Ariane Third Stage

The SPOT 1 Ariane third stage separated violently into fragments while over the eastern portion of Africa. The exact time and location of the

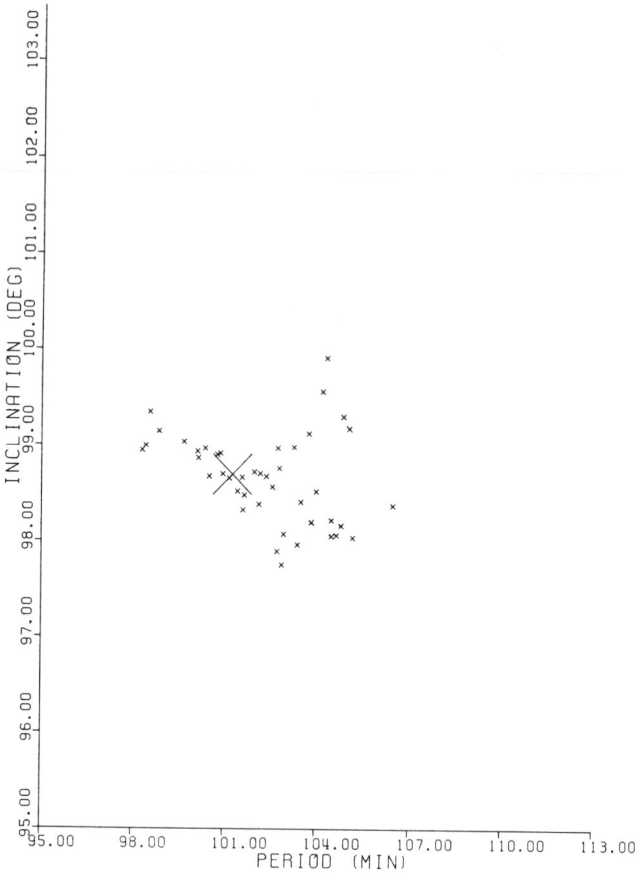

Fig. 3 Early data from the FPS-85 radar clearly shows an asymmetry in the retrograde debris.

breakup has been calculated by the Naval Space Surveillance System (NAVSPASUR) as 1939 and 23.2 s GMT at 7.0° N, 42.0° E (Appendix 3). Figure 1 indicates the groundtrack of the resulting debris cloud for 13½ h after the event. Although the debris cloud did not pass over the continental United States until more than 8 h later, personnel in the Space Surveillance Center (SSC) at the Cheyenne Mountain Complex in Colorado Springs, Colorado reported that the U.S. FPS-79 radar at Pirinclik, Turkey, noticed the debris within minutes of the fragmentation.

Monitoring of the SPOT 1 Ariane third stage prior to its fragmentation was sporadic. The European Space Agency (ESA) apparently lost telemetry contact with the rocket body very soon after it had successfully completed its mission. The NORAD space surveillance network was tasked to track the vehicle at a low priority befitting a spent rocket body. No NORAD data seems available that might shed light on the configuration or attitude of the Ariane third stage at the time of the event. But Teledyne Brown Engineering

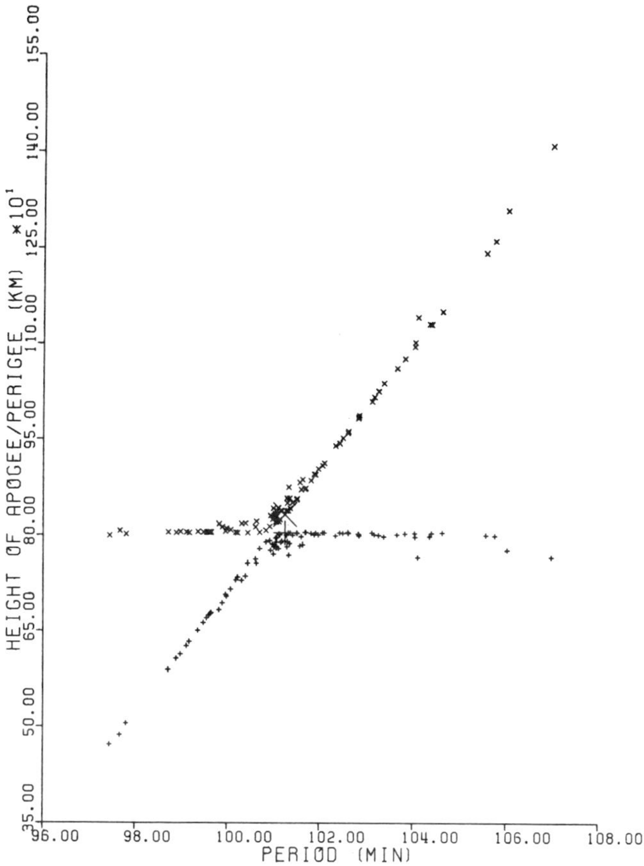

Fig. 4 Within three days of the breakup, 93 pieces of debris had been identified.

(TBE) found two other places that periodically monitored the Ariane third stage. Max White of the Royal Greenwich Observatory in England provided a summary of visual observations taken March–September 1986 (Appendix 8). A review of data collected at Edinburgh and in West Germany suggests that the rocket body was in a roll rather than an end-over-end tumble. The rate of this roll appears to have been increasing from May through September 1986. The last observation from Edinburgh was on September 7, and indicated regular optical variations at 30-s intervals.

In addition, Sven Grahn of the Swedish Space Corporation told the author an unidentified radar in West Germany had been tracking the Ariane third stage. According to him, the tumble period of the Ariane rocket body approximately one week before the event was 350 s. A lack of raw data prevented TBE from resolving the apparent discrepancy between the information supplied by White and Grahn. The possibility that the Ariane third stage did in fact experience a dramatic slowdown during September–

November should not be discounted, but the likelihood is assessed as low. However, the rocket body was almost certainly in a slow tumble (0.2-2 rpm) at the time of the event.

TBE requested and received from NORAD magnetic recording tapes made at the FPS-85 radar when the Ariane third stage debris cloud passed over that site for the first time about 8½ h after the breakup. As shown in Fig. 1, the pass geometry was excellent, with the debris entering in the center of coverage and exiting just to the southwest of the radar itself. Analysis of the FPS-85 tapes revealed 44 objects that could be correlated with the debris cloud. In retrospect, this relatively small number of debris objects detected may be attributed to the fact that the site was then configured for its primary role of missile warning.

Figure 2, a Gabbard diagram, indicates the apogee (×), perigee (+), and orbital period of the debris seen by the FPS-85 at this time. Orbital periods ranged from about 98 to 107 min, while altitude ranged between about 550 and 1300 km. At the time of passage, the actual altitude was close to 800 km. Of even greater interest was an apparent asymmetry in the distribution of debris in inclination and orbital period. Figure 3 clearly indicates that debris thrown in a retrograde (lower-orbital-period) direction mostly experienced increases in inclination, whereas posigrade (higher-or-

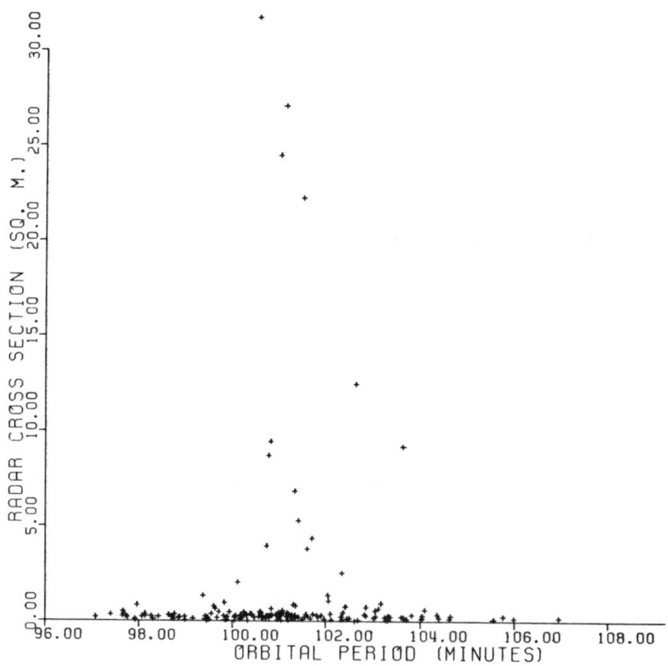

Fig. 5 Radar cross-section measurements made by the end of November 1986 indicated a relatively uniform distribution of fragments as a function of period.

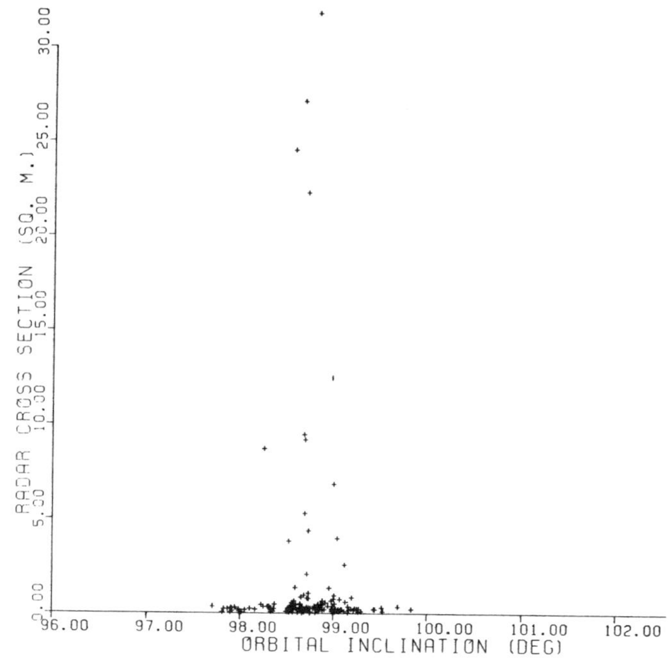

Fig. 6 Also apparent by the end of November 1986 was the uniform distribution of debris in RCS and inclination. Since the fragmentation took place close to the equator, inclination changes are easily related to the cross-track component of the ejection vector.

Fig. 7 By late February 1987, the debris had dispersed in a ring about the Earth approximately 30 deg wide at the equator.

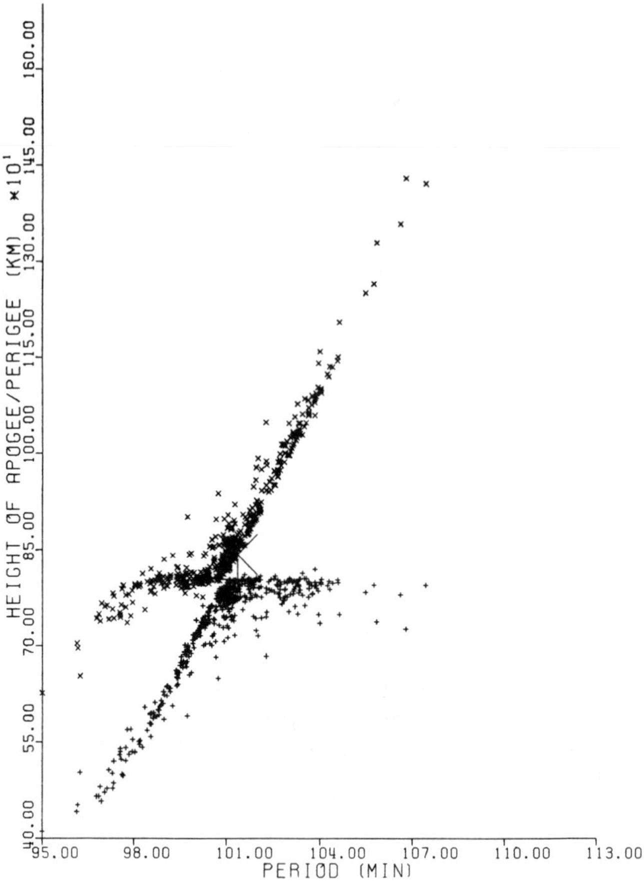

Fig. 8 Although some very-low-period debris has been affected by atmospheric drag, the have majority will remain in orbit for hundreds of years.

bital-period) debris experienced both increases and decreases in inclination. The retrograde debris was ejected rearward and to the left of the velocity vector. No useful RCS data were available on the data tapes provided to TBE.

Just minutes after leaving the FPS-85 coverage, the debris passed almost directly over the PARCS phased-array radar in North Dakota. However, since this pass was from south to north and the PARCS radar is optimally designed for southbound passes, TBE requested, but did not receive, data tapes from PARCS for the first good southbound pass, which occurred almost one full day after the breakup.

Although only a few dozen fragments had been seen by the FPS-85, almost immediately estimates from NAVSPASUR predicted more than 200 fragments. Within three days of the breakup, provisional element sets had been determined for 93 pieces of debris. The Gabbard diagram of Fig. 4

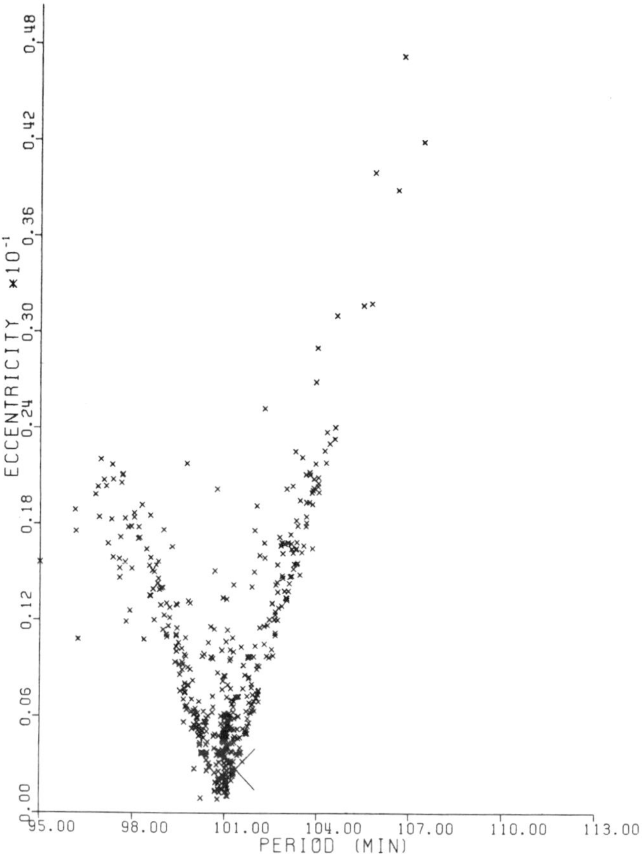

Fig. 9 The debris on the inside of the "V" diagram have experienced substantial radial ejection-velocities.

indicates the very ordered dispersion of this debris, with approximately equal numbers in the posigrade and retrograde directions. Atmospheric drag had yet to take its toll on the lower-period debris.

By the end of November, 274 pieces of debris had been tentatively associated with the third-stage fragmentation. More importantly, estimates of debris size were finally coming in from the FPS-85 radar. Figure 5 summarizes the 167 RCS values available at the end of November. In general, the fragments are evenly distributed around the original 101.25-min period of the rocket body prior to breakup, although the larger pieces appear slightly skewed toward higher orbital periods. A similar, relatively-uniform distribution can be seen in the RCS vs inclination diagram of Fig. 6. Due to the proximity of the event to the equator, the energy imparted in the cross-track (plane-change) direction is proportional to the sine of half the inclination-change angle.

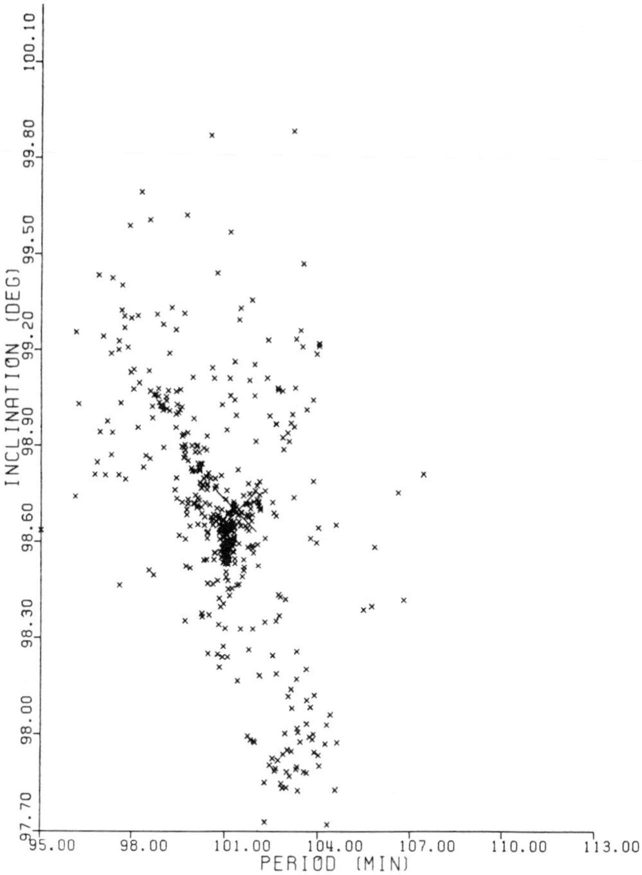

Fig. 10 From this diagram, the breakup seems to have consisted primarily of two jets emanating in opposite directions from the SPOT 1 rocket body.

Interestingly, four pieces of debris were assessed to have RCS values greater than 20 m^2—i.e., of the same order of magnitude as the Ariane rocket body before the fragmentation. This is not unexpected if the rocket body broke into several elongated pieces. The smallest RCS value registered was equivalent to a circular disc with a diameter of 8 cm. After a correction for Rayleigh scattering, however, the true physical dimension appears to be closer to 11 cm—close to the limiting observable size of a satellite at an altitude of 800 km.

Three months after the fragmentation, the magnitude of the incident was becoming clearer. Orbital parameters for a total of 465 fragments were being maintained by NAVSPASUR on a regular basis (Appendixes 1 and 2). Of these, 226 had been entered into the official NORAD Satellite Catalog. By June, 1989, pieces had been officially cataloged. Additional debris will almost certainly be discovered, but probably only in small quantities.

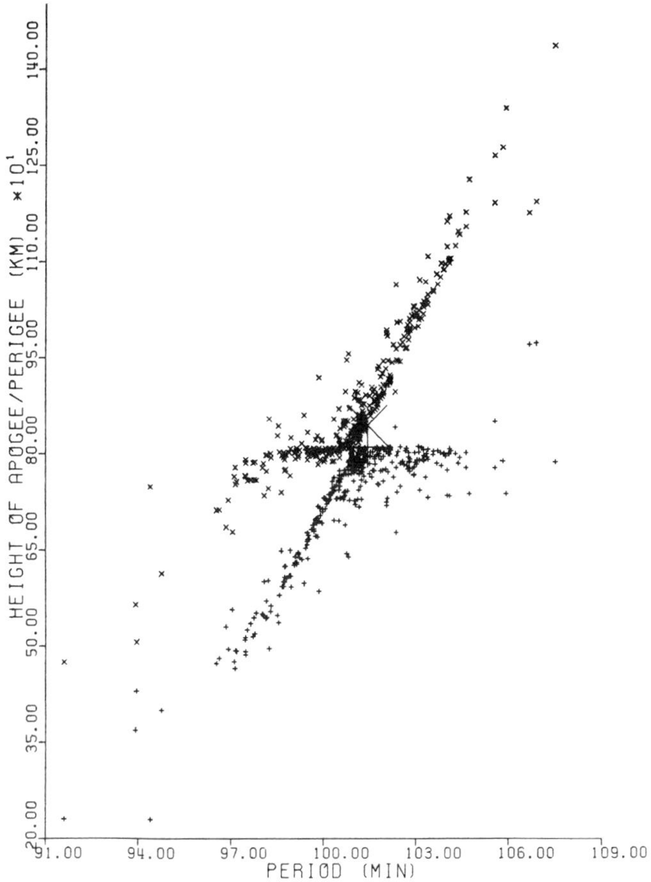

Fig. 11 Special FPS-85 observations failed to uncover a large number of "unknown" satellites or any new peculiarities in the debris cloud.

Due to the wide range in orbital periods, the debris are generally evenly distributed about the globe (Fig. 7), with the faster (lower-altitude) debris regularly overtaking the slower (higher-altitude). Also due to these period differences, the rates of precession of the nodes vary, resulting in a gradual separation of orbital planes. By late February 1987 this spread of right ascensions had grown to about 30 deg.

Figures 8–10 characterize the extent of the debris cloud in altitude, period, eccentricity, and inclination. Figure 8 confirms a high density of tracked debris in orbits near that of the original rocket body. A slight asymmetry toward retrograde orbits is also discernible. The very-low-period fragments are already being affected by atmospheric drag, as evidenced by decreasing apogees.

If the debris in this fragmentation had been directed only along the tangential velocity, the debris depicted in Fig. 9 would form a very distinct

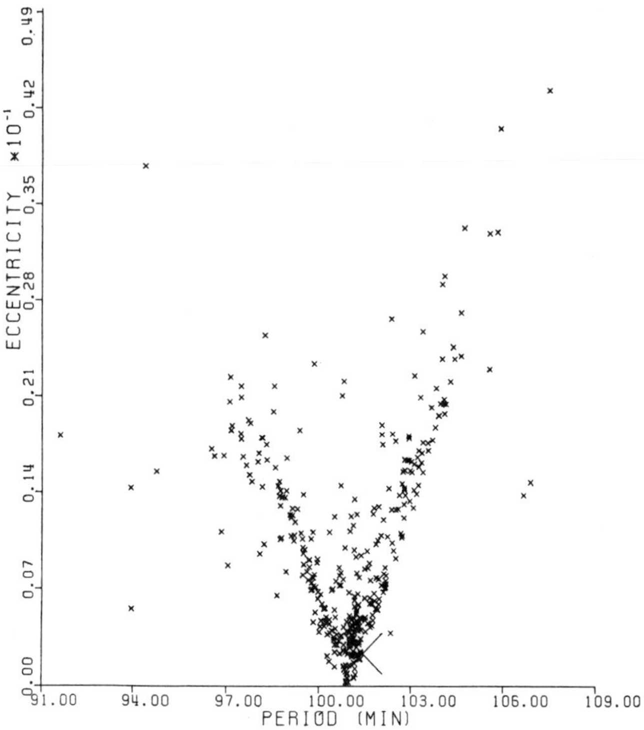

Fig. 12 This FPS-85 diagram is quite similar to the network assessment shown in Fig. 9.

Table 2 History of Ariane flights

Ariane flight	Ariane vehicle[a]	Date	Initial inclination, deg	Third-stage estimated lifetime (RAE), yr
1	1-01	Dec. 24, '79	17.6	1
2	1-02	May 23, '80	(failed)	
3	1-03	June 19, '81	10.5	20
4	1-04	Dec. 20, '81	10.5	8
5	1-05	Sept. 10, '82	(failed)	
6	1-06	June 16, '83	8.6	20
7	1-07	Oct. 19, '83	8.5	5
8	1-08	Mar. 5, '84	8.7	1
9	1-09	May 23, '84	6.7	5
10	2/3-01	Aug. 4, '84	7.0	50
11	2/3-02	Nov. 10, '84	7.0	120
12	2/3-03	Feb. 8, '85	7.1	50
13	2/3-04	May 8, '85	7.0	10
14	1-10	July 2, '85	7.0	50
15	2/3-05	Sept. 12, '85	(failed)	
16	1-11	Feb. 22, '86	98.7	200
17	2/3-06	Mar. 28, '86	7.1	100
18	2/3-07	May 30, '86	(failed)	

[a]The 1 refers to Ariane 1 configuration; 2/3 refers to Ariane 2 or 3 configuration. The third stage is lengthened for Ariane 2 and 3 missions.

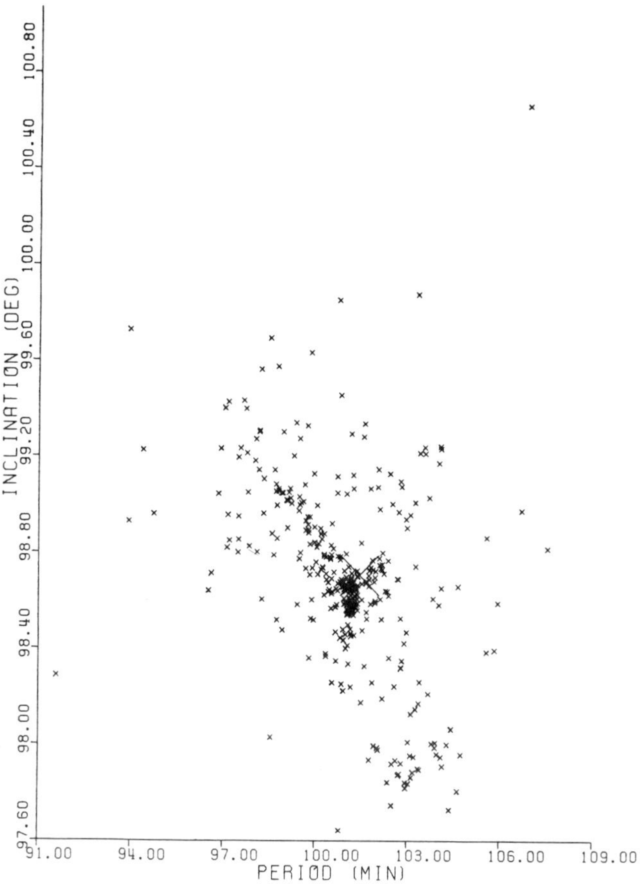

Fig. 13 The asymmetric relationship of inclination and period was verified by this FPS-85 "snapshot" of the debris cloud.

"V" shape. Fragments to the left of the "V" are being affected by atmospheric drag, while those inside the "V" have acquired substantial radial ejection-velocities. Figure 10 shows the best evidence yet of a marked asymmetry in the ejection of debris; the breakup seems to have been concentrated along two directions: one to the rear and left of the velocity vector and the other in the opposite direction, to the forward and right of the velocity vector.

On December 1, 1986, Donald Kessler of NASA Johnson Space Center requested U.S. Space Command to undertake special collection efforts using the FPS-85 and PARCS radars (Appendix 5). On January 15, 1987, the FPS-85 collected observations on the debris cloud in a pure satellite-tracking mode (i.e. without a missile warning fence erected). At the time, 80 pieces of debris had been officially cataloged and 379 more were being maintained as 8X,XXX satellites by NAVSPASUR. After processing the

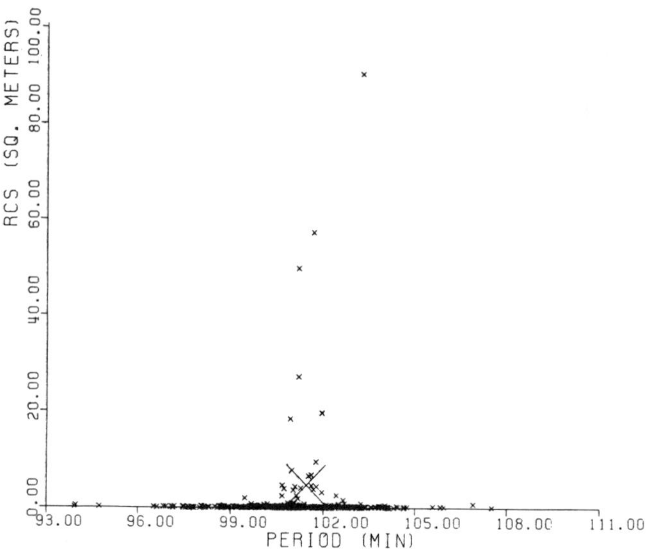

Fig. 14 This record of RCS vs period was made by the FPS-85 in January 1987.

FPS-85 data tapes, TBE discovered 43 cataloged objects, 301 8X,XXX objects, and 18 objects uncorrelated by the FPS-85. The total of 362 was only 79% of the "known" debris. Particularly puzzling was the absence of 37 (almost half) of the cataloged debris. The reason for these relatively poor statistics is not completely understood, but the aspect angle of some of the smaller pieces could have prevented an acceptable signal-to-noise ratio. Figures 11–13, derived from the FPS-85 data, are similar to Figs. 8–10, respectively, and show no gross discrepancies.

RCS values for the majority of the January 15, 1987 FPS-85 observations were available on the data tapes supplied to TBE. However, the accuracy of the values was restricted to 0.1 m^2. Summaries of these data are displayed in Figs. 14–19 as a function of period, eccentricity, and inclination. The diagrams are presented in pairs: the first with a scale of 0–160 m^2 and the second with a scale of 0–8 m^2 for better clarity of the smaller fragments. Overall, no significant features are apparent. In particular, Fig. 19 suggests a symmetric distribution as a function of inclination. During this pass over the FPS-85, the fragment retaining the designation of satellite 16615, the original Ariane third stage, was assessed to have a RCS of 4.7 m^2.

About two weeks later, on January 27, 1987, the debris cloud was observed by PARCS while in a special "small satellite" mode. In this condition software modifications attempt to remove the normal −30 dBSM sensitivity of the radar. Data tapes recorded for a 79-min period were provided to and reduced by TBE. Figures 20-28 are comparable to Figs. 11–19 for the FPS-85 and are constructed on the same scales. The most noticeable difference between the two sets of diagrams are the across-the-board smaller RCS values calculated by PARCS. This phenomenon was previously identified by TBE.[1] The normal −8 dBSM discrepancy between

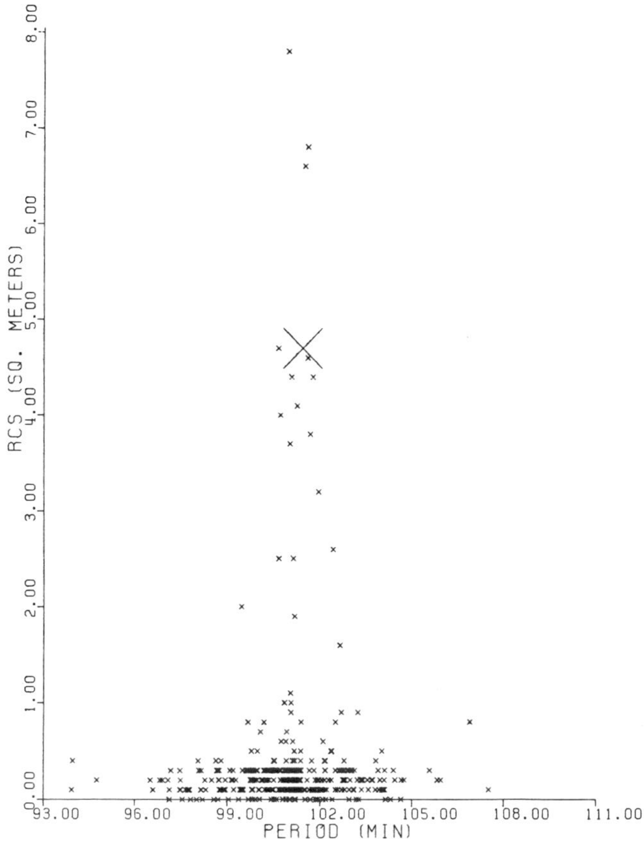

Fig. 15 This record of RCS vs period (expanded view) was made by the FPS-85 in January 1987. The granularity of the data (0.1 accuracy) causes the lines in the diagram.

PARCS and the FPS-85 becomes further accentuated when the small-satellite software patch is installed at PARCS. The values calculated by the FPS-85 are believed to be more accurate. In both cases, no attempt has been made to correct for Rayleigh scattering.

Previous Ariane-Third-Stage Anomalies

Ariane launches before the SPOT mission were confined to low-inclination, geostationary-transfer-type orbits (Table 2). The first Ariane mission was conducted on December 24, 1979 at an inclination of 17.6 deg. By 1983 it became apparent to TBE personnel that the Ariane third stage or its CAT 1 (Capsule Ariane Technologique) payload had been fragmented (Appendix 6). Although no debris were officially cataloged with the flight, many pieces of debris were seen "off and on" over a span of many years.

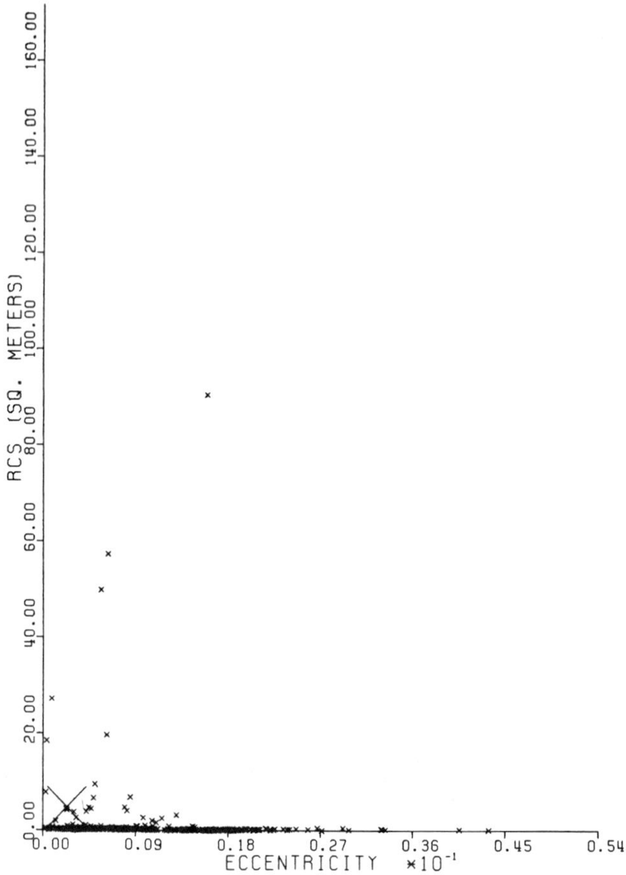

Fig. 16 RCS vs eccentricity from the FPS-85 in January 1987.

In April 1984 the author approached I. W. Pryke, head of the Washington, D.C. office of ESA, and asked if he had knowledge of the hypothesized breakup. The following month Pryke responded that ESA had been informed by the U.S. Millstone radar that the Ariane third stage had fragmented about 90 days after launch (Appendix 7). The causative agent was believed to be residual hydrogen and oxygen left onboard the third stage. The exact mechanism for the breakup was not offered, although Pryke noted that the hydrogen and oxygen valves were left open on subsequent flights.

A review of the February 19, 1987 NORAD satellite database revealed that no less than 21 8X,XXX numbers had been assigned to objects with characteristics similar to the first Ariane third stage during 1986. However, only two of these satellites were tracked recently, the rest having become "lost." At this late date it is impossible confidently to correlate

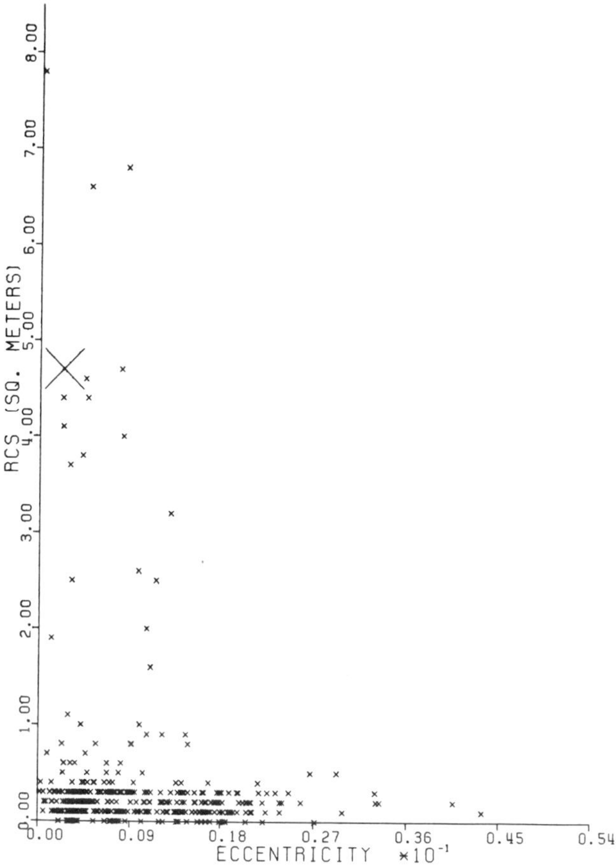

Fig. 17 RCS vs eccentricity (expanded view) from the FPS-85 in January 1987. The granularity of the data (0.1 accuracy) causes the lines in the diagram.

these debris to Ariane Flight 1. Solar-lunar perturbations and atmospheric drag should have combined to decay all Ariane Flight 1 debris by now. The possibility remains that this debris is of other origin, although this inclination is rarely used.

TBE reviewed all objects associated with Ariane flights (since Flight 1) that entered geostationary transfer orbits, i.e., all flights except the SPOT 1 mission. Table 3 gives the results. Ordinarily, only the third stage and the Sylda payload container (it being used) are cataloged. The Sylda is further identified by a very rapid decay, due in part to a high drag coefficient (area/mass). Several pieces of debris have been cataloged since the SPOT 1 Ariane third-stage breakup. This is due to recent special attention to Ariane rocket bodies. TBE has long noted the frequent appearance of uncataloged debris that can be associated with Ariane flights. The same February 19,

Table 3 Ariane deep-space rocket bodies and debris (as of March 1, 1987)

Ariane flight	International designator	Catalog number	Identity	Status
1	1979-104	11659	Third stage	Decayed Nov. 1982; numerous debris detected, not cataloged; see text
3	1981-57	12562	Third stage	In orbit
		14125	Debris	In orbit
4	1981-122	13025	Third stage	In orbit
6	1983-58	14130	Third stage	In orbit
		14151	Debris (Sylda?)	Decayed July 1982
		17302	Debris	In orbit
		17331	Debris	In orbit
7	1983-105	14423	Third stage	In orbit
8	1984-23	14787	Third stage	Lost; decayed?
9	1984-49	14989	Third stage	Decayed Mar. 1986
10	1984-81	15165	Third stage	In orbit
		15166	Debris (Sylda?)	In orbit; lost Sept. 1984; "found" Jan. 1985
11	1984-114	15388	Third stage	In orbit
		15389	Debris (Sylda?)	Decayed Oct. 1986
12	1985-15	15562	Third stage	In orbit
		15563	Debris (Sylda?)	Decayed Apr. 1986
13	1985-35	15679	Third stage	In orbit
		15680	Debris (Sylda?)	In orbit
14	1985-56	15876	Third stage	Lost Jan. 1986
		17255	Debris	In orbit
		17325	Debris	In orbit
		17332	Debris	In orbit
17	1986-26	16657	Third stage	In orbit
		16658	Debris (Sylda?)	Lost Nov. 1986; decayed?
		17253	Debris	In orbit
		17254	Debris	In orbit

1987 satellite database referred to above contained 35 satellites apparently related to Ariane flights. Of these only seven had been tracked recently.

A special examination of FPS-85 observations in September 1986 by TBE revealed seven objects with characteristics of Ariane flights during a brief observation interval. For the FPS-85 to detect these objects at the extreme ranges indicated, they must possess significant RCS values. Almost certainly objects less than 1 m^2 would go unseen. Therefore, if fragmentations like that experienced by the SPOT 1 Ariane third stage have occurred previously, very few of their debris would be detectable.

In Appendix 5 a request was made to conduct special observations of debris in Ariane-like orbits. TBE constructed fictitious element sets for hypothetical objects which might be related to a recent Ariane launch. Difficulties at the FPS-85 prevented a thorough search. A renewed effort is recommended.

Possible Causes of the Fragmentation

Possible causes of the fragmentation of the SPOT 1 Ariane third-stage will remain to an extent speculative. In general, the causes of satellite fragmentations have been attributed to explosions or collisions. Unfortunately, the reasons for about half of all fragmentations in space are still unknown. Furthermore, only two documented cases of collisions are available, and they were not completely representative of what might happen if a small, untracked object were to collide with a rocket body of the size of an Ariane third stage. Hence, the ability to distinguish between the effects

Fig. 18 RCS vs inclination from the FPS-85 in January 1987.

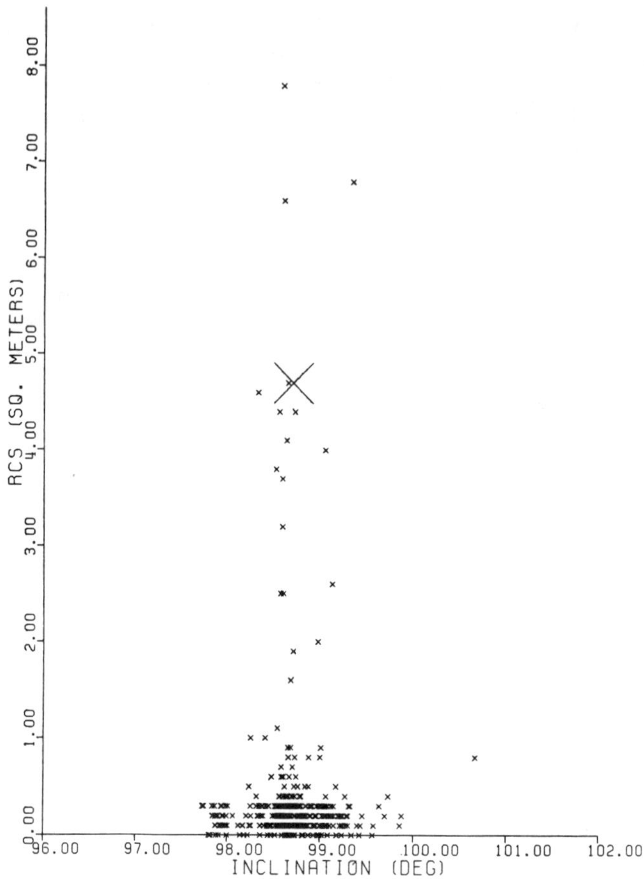

Fig. 19 RCS vs inclination (expanded view) from the FPS-85 in January 1987. The granularity of the data (0.1 accuracy) causes the lines in the diagram.

of an explosion and of an accidental collision is still in the formative stage. Promising techniques for this purpose are now being evaluated, but none have yet been validated.

Familiarity with the history of satellite breakups points to similarities between the SPOT 1 Ariane third stage breakup and the fragmentations that plagued U.S. Delta second stages for several years. The Delta second stages also produced intense breakups with hundreds of detectable pieces, and some occurred as long as three years after launch. Perhaps more importantly, a TBE report in 1985 identified critical parameters common to the exploding Delta second stages[2]: a failure to deplete residual propellants and insertion into a Sun-synchronous orbit with a descending node of 1000 local Sun time. These parameters are also common to the Ariane third stage in question.

Although the exact mechanlsm of the Delta second-stage breakups was never definitely pin-pointed, two hypotheses are thought credible. Repeated expansion and contraction of the Delta second stage may have ruptured the common bulkhead between the residual fuel and oxidizer. Since the Delta second stage used hypergolic propellants, simple contact could cause an explosion. The Ariane third stage has a structural design similar to the Delta second stage, but its propellants are not hypergolic under normal conditions.

The second hypothesis called for a gradual heating of the rocket body and over-pressurization and eventual rupture of propellant tanks due to structural fatigue. This scenario is potentially applicable to the Ariane third stage; but the stage has pressure-relief valves designed to open at 3 bars (hydrogen tank) and 2 bars (oxygen tank). These valves might become stuck

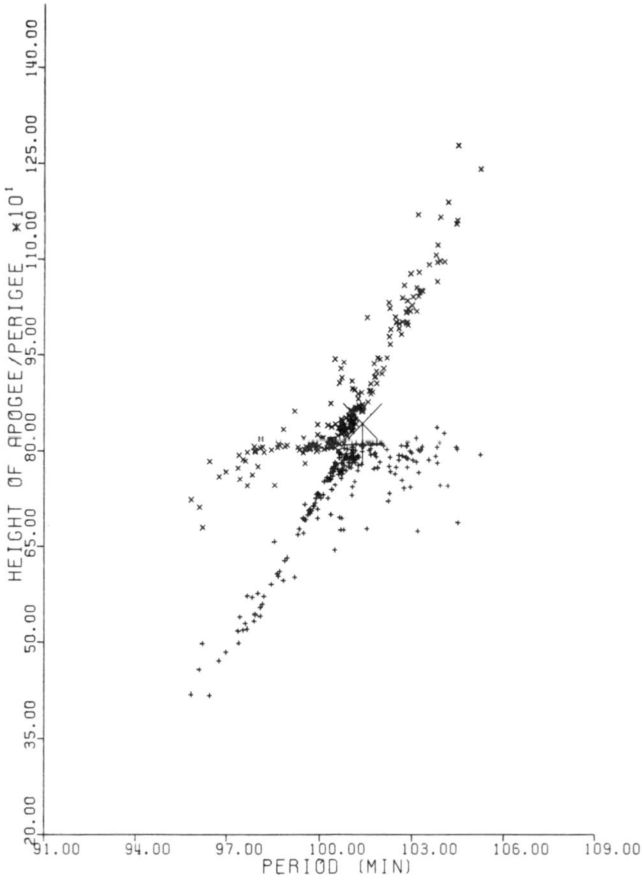

Fig. 20 Gabbard diagram derived from special PARCS observations in January 1987.

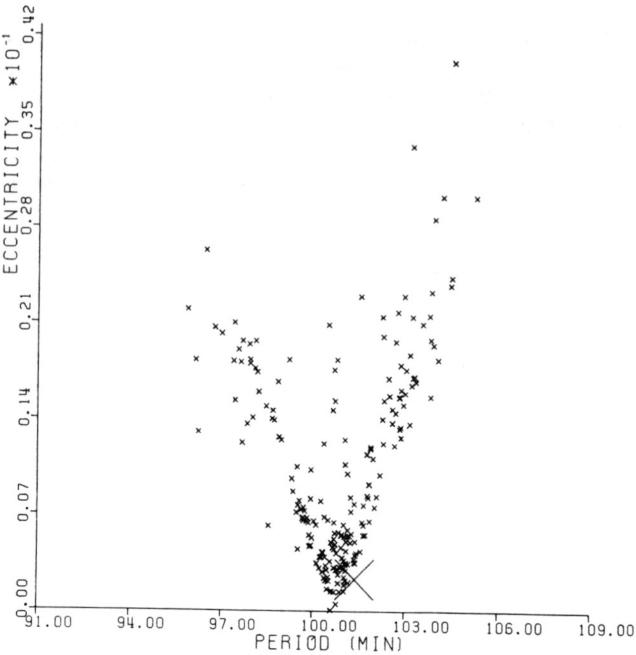

Fig. 21 Eccentricity vs period from PARCS in January 1987.

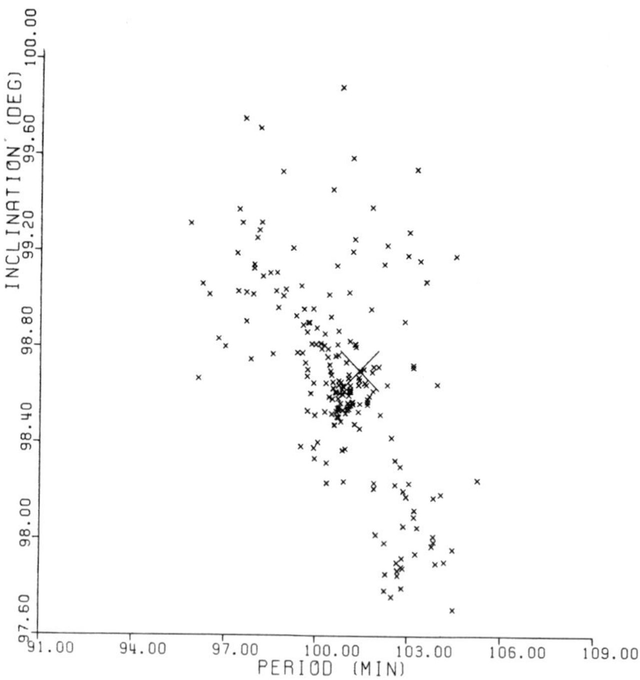

Fig. 22 Inclination vs period from PARCS in January 1987.

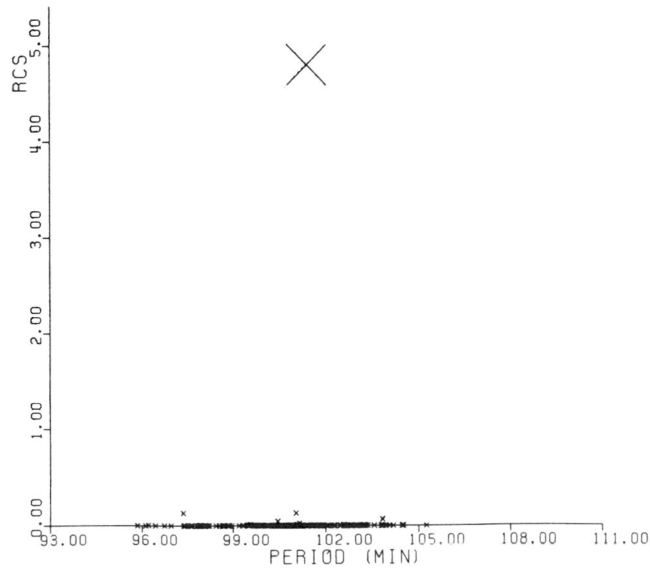

Fig. 23 RCS vs period from PARCS in January 1987.

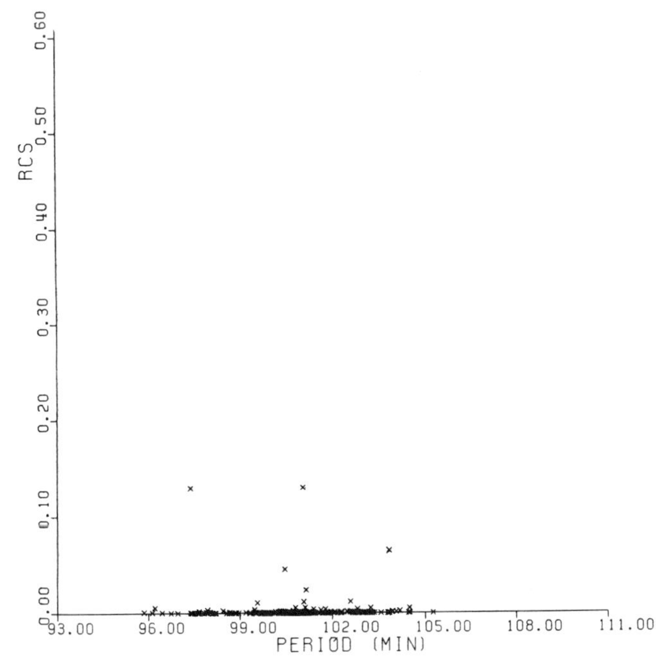

Fig. 24 RCS vs period (expanded view) from PARCS in January 1987.

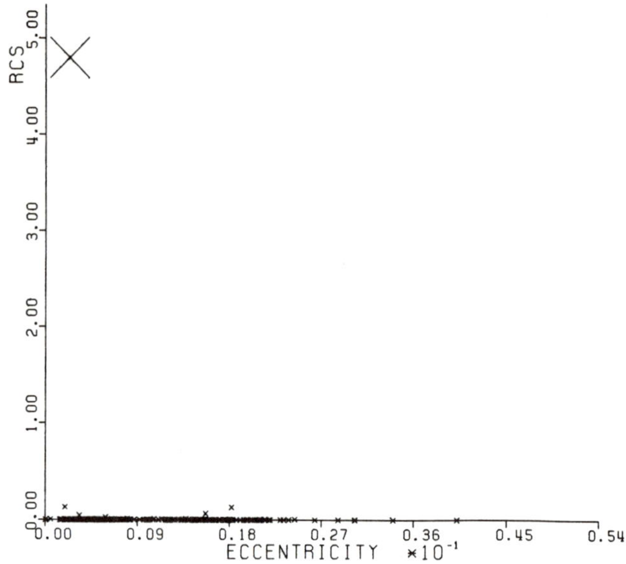

Fig. 25 RCS vs eccentricity from PARCS in January 1987.

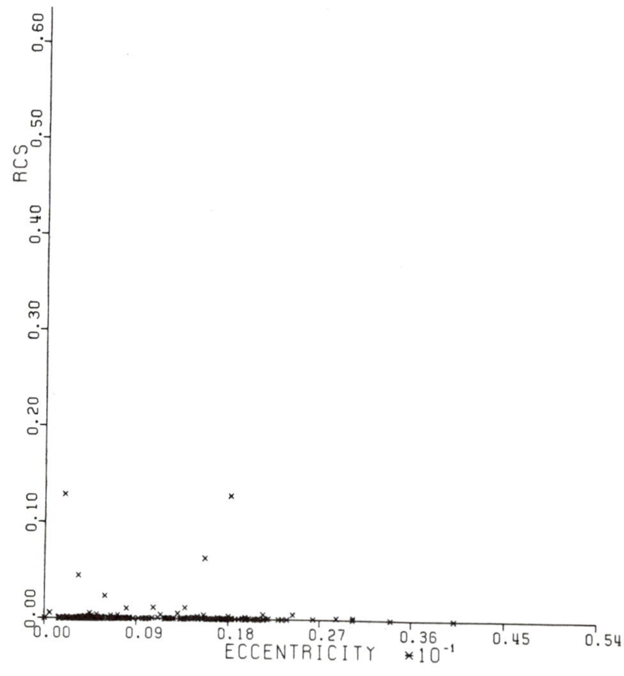

Fig. 26 RCS vs eccentricity (expanded view) from PARCS in January 1987.

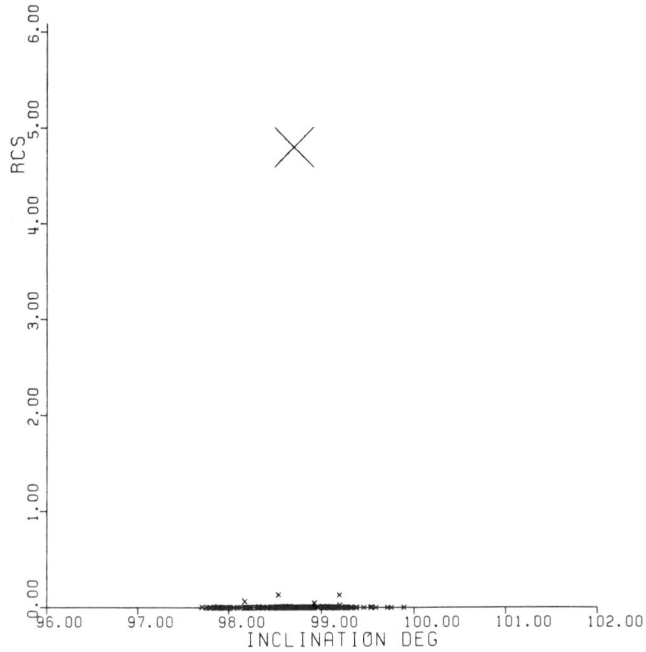

Fig. 27 RCS vs inclination from PARCS in January 1987.

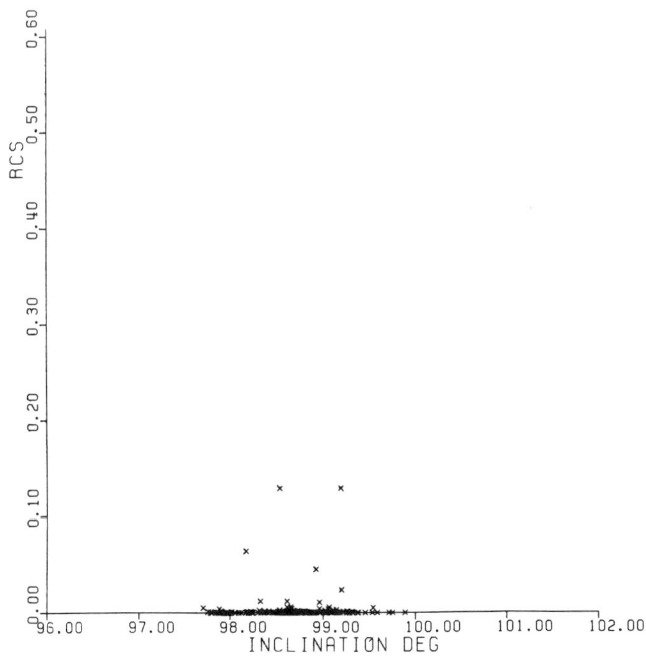

Fig. 28 RCS vs inclination (expanded view) from PARCS in January 1987.

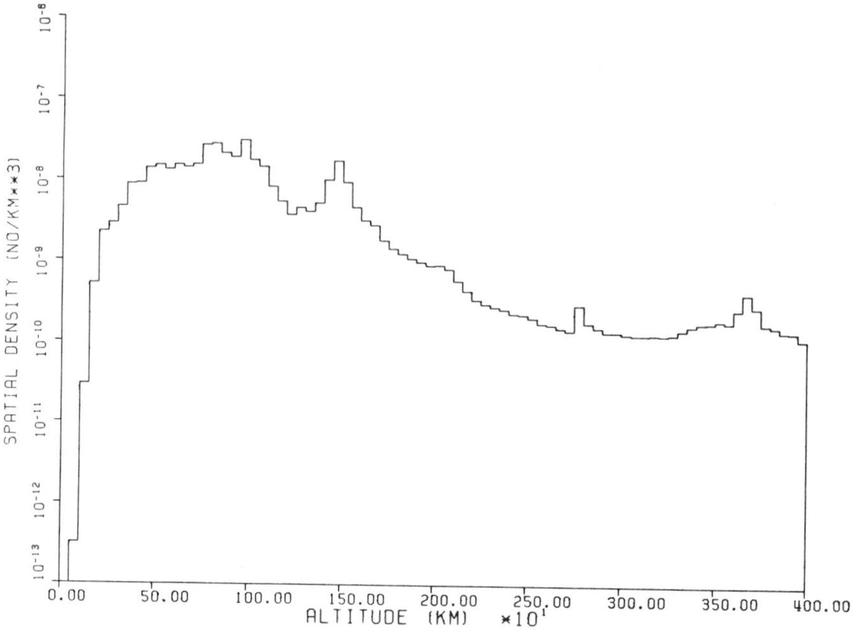

Fig. 29 Near-Earth spatial density calculated in February 1987. Actual values for the 750-900-km regime are actually slightly higher than shown due to a database deficiency.

after several months in space. The implied over-pressurization of the first Ariane third stage (previously discussed) forces a similar dilemma.

The case for an accidental collision of the SPOT 1 Ariane third stage with another object suffers from the extremely low probability of such an event. None of the 6,000+ objects tracked by NORAD in November 1986 were responsible for the breakup. The probability of the SPOT 1 rocket body being struck by a tracked object was only once in more than 13,000 years. Although the assumed small satellite population is many times that of the known satellite population, the probability of collision is still low.

Theoretically, the number of debris of a given size can be useful in distinguishing between a collision and an explosion.[3] However, the smallest particles seen from the Ariane third stage are only about 0.01 m^2 in true cross section. This can be equated to a mass of approximately 340 gm. In the case of an explosion, the number of debris with this mass or greater should be on the order of 600 [Ref. 3, Eqs. (1a) and (1b)]. On the other hand, a catastrophic collision should result in about only 250 pieces of mass 340 g or greater if the impacting particle was of sufficient size to induce total fragmentation [Ref. 3, Eq. (5)]. Moreover, the size of the particle would have to have been large enough to be readily detectable by the NORAD Space Surveillance Network [Ref. 3, Eq. (4)]. Due to NORAD operational constraints and the assumptions used in the above calculations, the difference between 600 and 250 pieces of debris for the two different

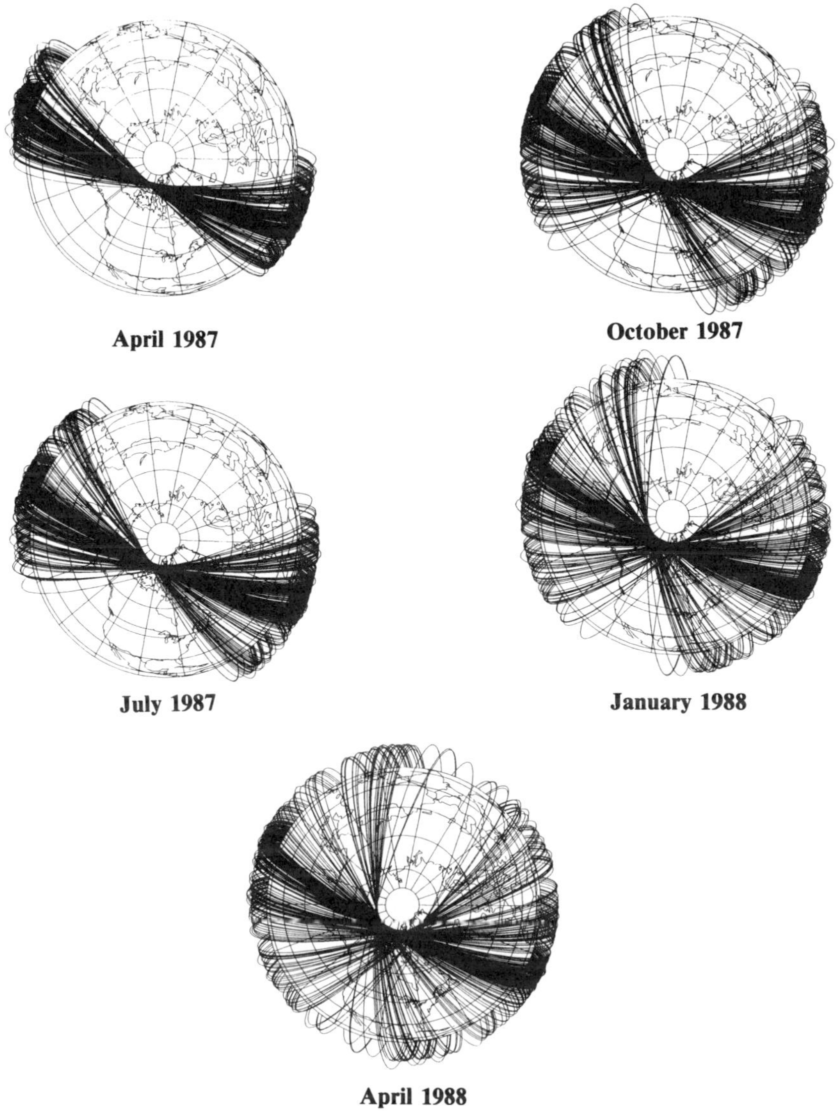

Fig. 30 SPOT 1 R/B debris planes.

mechanisms is perhaps statistically insignificant. If a selection between an explosion and a collision had to be made with the evidence now available, an explosion would be thought to be more likely. The source of this explosion is, of course, unknown. The primary candidates are over-pressurization, ignition of residual propellants, and detonation of the on-board destruct devices. Radar imaging of some of the larger pieces of debris might provide additional evidence that could narrow the potential scenarios.

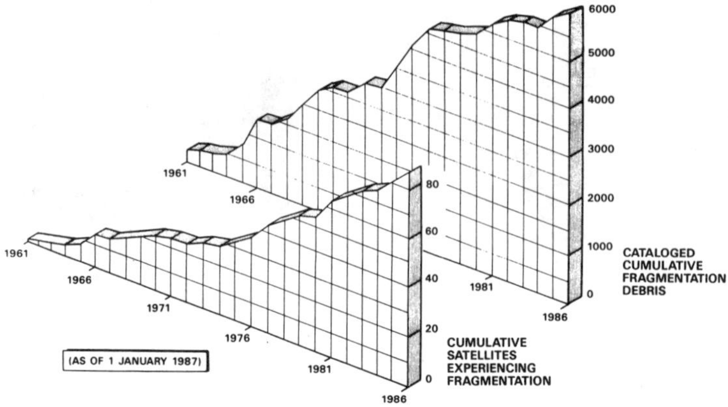

Fig. 31 More than a third of the almost 18,000 orbiting objects cataloged since the launch of Sputnik 1 in 1957 have originated in satellite fragmentations. Disturbingly, the frequency of fragmentations has increased during the 1980's.

Long-Term Effects on the Near-Earth Environment

Due to the high altitude of the SPOT 1 Ariane third stage when it broke up, the majority of the debris were injected into very-long-lived orbits. As noted earlier, the altitude regime now affected ranges from about 500 to 1400 km. Of particular concern is the regime between 750 and 900 km. As the result of this single event, the spatial density (average number of satellites in a given volume of space) for this region doubled between February 1986 and February 1987. Figure 29 depicts the spatial density calculated from a NORAD satellite database (cataloged plus provisional satellites) in mid-February 1987. The spatial density in the 750-900-km region is actually higher than shown because the NORAD database used did not include about 100 provisional satellites being maintained by NAVSPASUR. When all new trackable debris are accounted for, the region between 800 and 850 km should be the most densely populated band about the Earth.

Since the inclination of the SPOT 1 Ariane third stage was nearly polar, the resulting debris will traverse virtually all latitudes. Hence, most Earth satellites are substantially affected by this new debris "cloud." Figure 30 illustrates the dispersion of the debris orbit planes at three-month intervals. The plane extremes are separating at the rate of about 10 deg per month, implying a near-uniform distribution about the Earth will occur by mid-1988. A beneficial side-effect, this process eliminates concentrations of satellites that now result in regions of spatial density much above those shown in Fig. 29.

The remaining figures in this section (Figs. 31–33) summarize the state of the Earth-satellite population and the adverse effect that satellite breakups are having on the environment. Although it was one of the worst known breakups, the SPOT 1 Ariane third-stage fragmentation is certainly not an

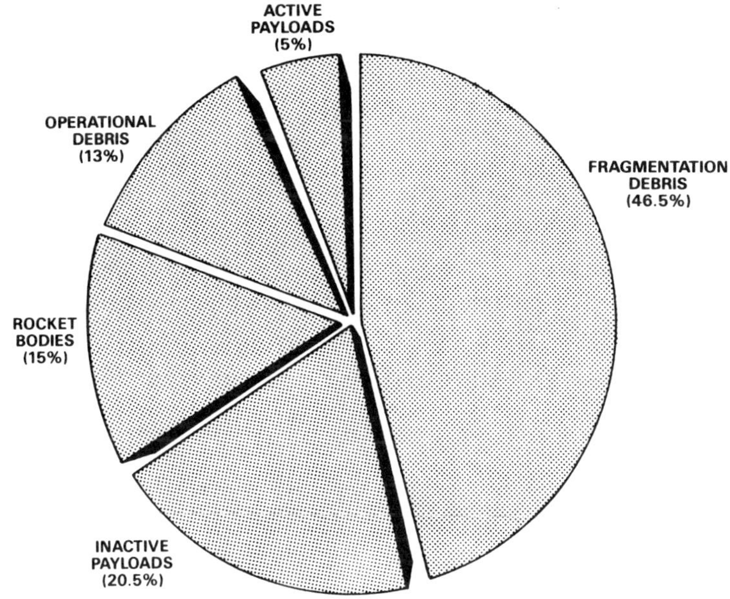

Fig. 32 Of all the man-made objects circling the Earth on January 1, 1987, only 5% were operational payloads.

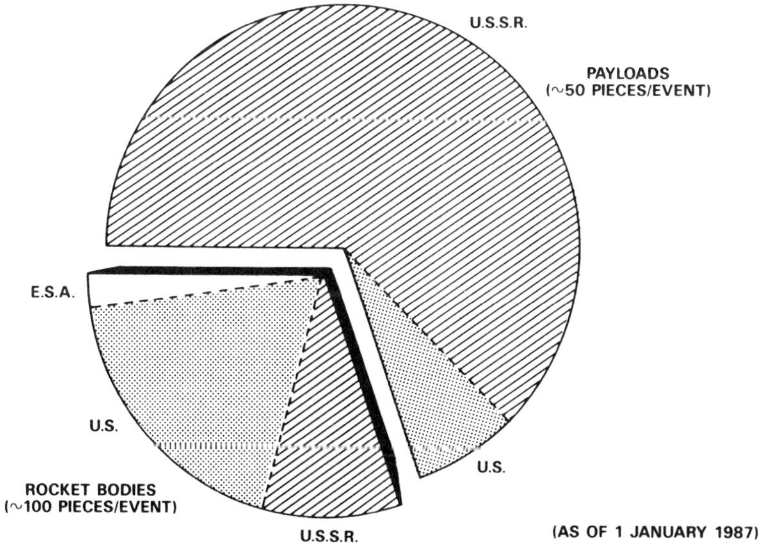

Fig. 33 Satellite payloads are more likely to be the subject of fragmentations, but on the average more trackable debris are produced by rocket-body explosions.

uncommon event. If the growing space-debris hazard is to be curtailed, actions must be taken to correct design defects or operational procedures that can create additional space debris.

References

[1]Kling, R., "Postmortem of a Hypervelocity Impact," Teledyne Brown Engineering, Colorado Springs, CO, Rept. CS86-LKD-001, Sept. 1986.

[2]Gabbard, J. R., "A Later Look at Delta Second Stage On-orbit Explosions," Teledyne Brown Engineering, Colorado Springs, CO, Rept. CS85-BMDSC-00-24, March 1985.

[3]Su, S.-Y. and Kessler, D. J., "Contribution of Explosion and Future Collision Fragments to the Orbital Debris Environment," *Advances in Space Research,* Vol. 5, No. 2, 1985, pp. 25–34.

Appendix 1
SPOT 1 R/B Cataloged Debris, February 25, 1987

```
1 16615U 86019  C 87055.99505246  .00000288  00000-0  00000-0 0 01572
2 16615 098.6950 132.1570 0027139 082.6636 277.8178 14.20713967052636
1 17110U 86019  E 87055.93744386  .00001066  00000-0  00000-0 0 00335
2 17110 098.9940 133.0921 0096341 078.5519 282.6989 14.05981590014470
1 17111U 86019  F 87055.95211533 -.00013050  00000-0  00000-0 0 00608
2 17111 099.0098 141.1923 0127483 240.1856 118.7192 14.51902528014955
1 17112U 86019  G 87055.93629152  .00000475  00000-0  00000-0 0 00423
2 17112 098.7226 136.6544 0088890 235.4271 123.9083 14.45823126014887
1 17113U 86019  H 87055.98017873  .00013898  00000-0  00000-0 0 00591
2 17113 098.5225 133.7097 0097121 261.2270 097.8488 14.44236497014863
1 17114U 86019  J 87055.93712989  .00004274  00000-0  00000-0 0 00292
2 17114 098.5154 133.3346 0089094 263.1576 096.0043 14.42163837014849
1 17115U 86019  K 87055.97997586 -.00000260  00000-0  00000-0 0 00466
2 17115 098.8302 137.2049 0062683 260.1298 099.3379 14.41532013014847
1 17116U 86019  L 87055.92924291  .00000079  00000-0  00000-0 0 00440
2 17116 099.1124 134.6082 0097052 061.2054 299.9356 14.06955840014481
1 17117U 86019  M 87055.96678277  .00000368  00000-0  00000-0 0 00345
2 17117 098.9986 131.4823 0147011 071.4400 290.3198 13.95877629014374
1 17118U 86019  N 87055.96827362  .00000106  00000-0  00000-0 0 00246
2 17118 099.2124 131.9315 0208840 073.8117 288.6429 13.84191579014258
1 17119U 86019  P 87055.98437256  .00000579  00000-0  00000-0 0 00512
2 17119 098.9565 140.1182 0103321 243.0744 116.0439 14.48860183014927
1 17120U 86019  Q 87055.97019328  .00001153  00000-0  00000-0 0 00787
2 17120 099.0575 134.9079 0068945 058.7900 302.0548 14.12491258014545
1 17121U 86019  R 87055.96557019  .00000292  00000-0  00000-0 0 00289
2 17121 098.7876 131.5281 0075849 066.1843 294.7800 14.10435429014526
1 17122U 86019  S 87055.96327753  .00000016  00000-0  00000-0 0 00217
2 17122 098.6519 132.1606 0013715 104.3178 256.0080 14.24043961014665
1 17123U 86019  T 87055.95639893  .00000106  00000-0  00000-0 0 00410
2 17123 098.5088 129.3613 0051934 107.3178 253.4247 14.17429207014591
1 17124U 86019  U 87055.94517938  .00000564  00000-0  00000-0 0 00176
2 17124 098.6445 131.6155 0024829 078.8667 281.5854 14.21457681052641
1 17126U 86019  V 87055.96068131  .00713398  00000-0  00000-0 0 00408
2 17126 098.6353 143.7075 0156166 215.6950 143.4373 15.15670682053557
1 17127U 86019  W 87055.99868154  .00051852  00000-0  00000-0 0 00786
2 17127 099.1873 148.1550 0182592 227.2376 131.3944 14.80006843053228
1 17128U 86019  X 87055.93299128  .00001195  00000-0  00000-0 0 00235
2 17128 099.0135 130.5453 0178257 075.7948 286.3464 13.89669738014300
1 17129U 86019  Y 87055.99252447  .00000208  00000-0  00000-0 0 00239
2 17129 098.6910 131.7918 0036423 082.6448 277.9417 14.18787765014618
1 17130U 86019  Z 87055.98224746  .00000635  00000-0  00000-0 0 00370
2 17130 098.7165 131.6157 0050012 074.0932 286.6295 14.16059094014582
1 17131U 86019 AA 87055.98624993  .00000304  00000-0  00000-0 0 00282
2 17131 098.5547 131.3116 0033514 149.4233 210.9468 14.25698083014686
1 17132U 86019 AB 87055.93988615  .00035205  00000-0  00000-0 0 00639
2 17132 098.6921 131.2211 0102912 137.0124 223.9701 14.18466480014590
1 17133U 86019 AC 87055.97493449  .00005463  00000-0  00000-0 0 00644
2 17133 099.0779 142.9478 0145463 233.2532 125.5818 14.57036678015009
1 17150U 86019 AD 87055.98432152  .00001746  00000-0  00000-0 0 00211
2 17150 098.5451 131.1559 0036114 145.5076 214.9016 14.25891619014583
1 17151U 86019 AE 87055.94388656  .00000993  00000-0  00000-0 0 00401
2 17151 098.5737 131.2646 0051942 148.6442 211.8410 14.25380562014579
1 17152U 86019 AF 87055.99072444  .00000055  00000-0  00000-0 0 00298
2 17152 098.7247 131.1981 0064369 079.7732 281.1238 14.12975508014453
1 17153U 86019 AG 87055.99169885  .00000240  00000-0  00000-0 0 00178
2 17153 098.7094 132.3252 0027313 080.2689 280.2125 14.20752964014503
1 17154U 86019 AH 87055.94912766 -.00001204  00000-0  00000-0 0 00338
2 17154 098.7226 129.7765 0101498 092.8405 268.4924 14.04692693014365
1 17155U 86019 AJ 87055.93270336  .00000003  00000-0  00000-0 0 00434
2 17155 099.0718 132.9089 0128359 064.8347 296.6599 14.00242265014318
1 17156U 86019 AK 87055.94713932 -.00001397  00000-0  00000-0 0 00427
2 17156 098.4884 129.2772 0037554 062.9120 297.6431 14.18239793014506
1 17157U 86019 AL 87055.96658722  .00002794  00000-0  00000-0 0 00370
2 17157 098.6049 131.5439 0060701 140.5969 220.0206 14.25284189014532
1 17183U 86019 AM 87055.93040528  .00002672  00000-0  00000-0 0 00404
2 17183 098.6098 125.6143 0211976 083.4796 279.0969 13.88039047014153
1 17184U 86019 AN 87055.93320383 -.00000444  00000-0  00000-0 0 00513
2 17184 098.9695 132.2928 0111524 089.8434 271.6047 14.02980977014308
1 17185U 86019 AP 87055.93554750  .00005076  00000-0  00000-0 0 00415
2 17185 099.0120 140.2980 0085757 229.8249 129.5983 14.47126202014750
1 17186U 86019 AQ 87055.93689415 -.00009431  00000-0  00000-0 0 00266
2 17186 098.6244 131.7050 0061239 153.9259 206.5584 14.26173463014541
```

continued...

Appendix 1 (cont'd)
SPOT 1 R/B Cataloged Debris, February 25, 1987

```
1 17187U 86019 AR 87055.93228798 -.00000266  00000-0  00000-0 0 00357
2 17187 099.0040 140.3103 0105606 243.8553 115.2319 14.47688356014692
1 17188U 86019 AS 87055.94961332  .00004988  00000-0  00000-0 0 00463
2 17188 099.1884 131.3913 0268722 053.4630 309.1461 13.85291194014186
1 17189U 86019 AT 87055.97386484  .00001324  00000-0  00000-0 0 00366
2 17189 098.0587 118.3808 0230159 080.1627 282.5950 13.79153260014180
1 17190U 86019 AU 87055.94540132  .00003098  00000-0  00000-0 0 00378
2 17190 098.8969 138.6297 0093567 250.1093 109.0570 14.45960571014745
1 17193U 86019 AV 87055.96782769  .00002840  00000-0  00000-0 0 00416
2 17193 099.0405 141.9955 0130078 238.8624 120.0339 14.53455867014973
1 17194U 86019 AW 87055.98304388  .00004319  00000-0  00000-0 0 00284
2 17194 098.7309 135.3356 0047708 246.2705 113.4040 14.37849707014804
1 17195U 86019 AX 87055.98437930  .00000657  00000-0  00000-0 0 00196
2 17195 098.5426 132.6023 0058609 190.6895 169.3614 14.34495180014770
1 17196U 86019 AY 87053.94758621  .00008531  00000-0  00000-0 0 00328
2 17196 098.6213 129.7319 0051538 110.8311 249.8953 14.25381715014376
1 17197U 86019 AZ 87055.93183048  .00003523  00000-0  00000-0 0 00387
2 17197 098.6784 130.7564 0096771 066.8959 294.2923 14.15204877014463
1 17198U 86019 BA 87055.99513589  .00000211  00000-0  00000-0 0 00203
2 17198 098.9721 131.2969 0146645 077.9368 283.8715 13.96469802014288
1 17199U 86019 BB 87055.99377548  .00000112  00000-0  00000-0 0 00331
2 17199 098.6812 129.1343 0110119 082.5539 278.8663 14.03101609014359
1 17200U 86019 BC 87055.94912534  .00000849  00000-0  00000-0 0 00167
2 17200 098.5430 132.0125 0047342 180.3710 179.8008 14.31106168014633
1 17201U 86019 BD 87055.94026331  .00000863  00000-0  00000-0 0 00390
2 17201 099.1643 137.7033 0047094 012.8497 347.4417 14.21268921014533
1 17202U 86019 BE 87055.96812017  .00011400  00000-0  00000-0 0 00149
2 17202 098.3503 124.6424 0123070 109.5899 251.9144 14.02952967014443
1 17203U 86019 BF 87055.98501709  .00000325  00000-0  00000-0 0 00217
2 17203 099.0814 132.2685 0154429 070.4690 291.3605 13.94623229014264
1 17204U 86019 BG 87055.92936292  .00003903  00000-0  00000-0 0 00193
2 17204 098.1715 121.6282 0164427 101.7826 260.2367 13.93914541014256
1 17205U 86019 BH 87055.95866961  .00007547  00000-0  00000-0 0 00459
2 17205 098.3678 124.7971 0127163 102.8457 258.7491 14.01517646014327
1 17206U 86019 BJ 87055.97803132  .00002038  00000-0  00000-0 0 00307
2 17206 098.6826 131.8348 0035666 108.8248 251.7355 14.20072754014524
1 17207U 86019 BK 87055.98687548  .00006895  00000-0  00000-0 0 00394
2 17207 098.5774 131.0175 0055210 141.1620 219.4104 14.24288926014579
1 17208U 86019 BL 87055.96468575 -.00004705  00000-0  00000-0 0 00209
2 17208 098.9955 135.6834 0071803 359.4107 000.7525 14.20969428014530
1 17209U 86019 BM 87055.97002415  .00000899  00000-0  00000-0 0 00207
2 17209 098.5419 132.7882 0062742 193.6329 166.3731 14.35686551014682
1 17210U 86019 BN 87055.96312754 -.00001007  00000-0  00000-0 0 00201
2 17210 098.8281 136.3899 0054309 256.8261 102.7429 14.37740953014709
1 17211U 86019 BP 87055.94985872  .00001808  00000-0  00000-0 0 00283
2 17211 098.7766 135.1813 0035533 244.1585 115.6497 14.34224787014670
1 17212U 86019 BQ 87055.99728780  .00000080  00000-0  00000-0 0 00205
2 17212 098.7256 130.9071 0073323 083.5525 277.4534 14.10966636014435
1 17219U 86019 BR 87055.96176813  .00000323  00000-0  00000-0 0 00168
2 17219 098.3274 126.6562 0066755 093.2316 267.7047 14.13406596014564
1 17220U 86019 BS 87055.93695454  .00014607  00000-0  00000-0 0 00412
2 17220 098.6264 132.0622 0047617 107.9901 252.7033 14.28497894014583
1 17221U 86019 BT 87055.94122604  .00001506  00000-0  00000-0 0 00263
2 17221 098.6673 132.6924 0051588 150.4532 210.0137 14.26591698014581
1 17222U 86019 BU 87055.95082648 -.00001242  00000-0  00000-0 0 00233
2 17222 098.6522 124.4715 0240333 096.6097 266.2989 13.76911431014041
1 17223U 86019 BV 87055.98257111  .00000540  00000-0  00000-0 0 00190
2 17223 098.5605 131.5166 0039055 156.7532 203.5985 14.26742543014559
1 17224U 86019 BW 87055.97759021  .00003104  00000-0  00000-0 0 00220
2 17224 098.7449 135.8679 0064464 266.6733 092.7648 14.39671566014688
1 17225U 86019 BX 87055.96013433  .00028422  00000-0  00000-0 0 00343
2 17225 098.9405 145.0837 0207678 236.4995 121.6770 14.79280465015073
1 17226U 86019 BY 87055.97842135  .00002154  00000-0  00000-0 0 00270
2 17226 098.5615 131.1637 0018038 053.0953 307.2433 14.24874997014463
1 17227U 86019 BZ 87055.99373986  .00002423  00000-0  00000-0 0 00253
2 17227 098.8941 138.6270 0077344 245.0229 114.3486 14.45050743014663
1 17228U 86019 CA 87055.96592192  .00000325  00000-0  00000-0 0 00320
2 17228 098.6688 132.8298 0014267 332.7013 027.3975 14.26975732014485
1 17229U 86019 CB 87055.95106238  .00002379  00000-0  00000-0 0 00200
2 17229 098.6795 133.3893 0013543 193.7923 166.3451 14.30156273014550
1 17230U 86019 CC 87055.95539617  .00004624  00000-0  00000-0 0 00309
2 17230 098.5033 130.5037 0035146 342.7466 017.3116 14.25570065014530
1 17231U 86019 CD 87055.98502863  .00007439  00000-0  00000-0 0 00561
2 17231 099.0604 143.0313 0153999 226.0869 132.8115 14.58337287014811
1 17232U 86019 CE 87055.96449228  .00006352  00000-0  00000-0 0 00300
2 17232 099.0109 141.6511 0113594 221.9130 137.3907 14.54915257014733
```

continued...

Appendix 1 (cont'd)

SPOT 1 R/B Cataloged Debris, February 25, 1987

```
1 17233U 86019 CF 87055.93511850  .00000247  00000-0  00000-0 0 00160
2 17233 098.5312 130.2911 0043381 126.4260 234.1484 14.21546992014467
1 17234U 86019 CG 87055.97182991  .00019798  00000-0  00000-0 0 00380
2 17234 098.6070 134.4795 0065618 248.7542 110.7204 14.44350857014630
1 17235U 86019 CH 87055.97842538  .00011400  00000-0  00000-0 0 00349
2 17235 098.5596 131.1639 0019777 042.0199 318.3051 14.24882167014539
1 17236U 86019 CJ 87055.99761312  .00002932  00000-0  00000-0 0 00207
2 17236 098.5785 131.1463 0051591 121.6057 239.0728 14.23832148014423
1 17237U 86019 CK 87055.97874432  .00002068  00000-0  00000-0 0 00237
2 17237 098.3283 128.6050 0027696 355.5660 004.5830 14.25740075014568
1 17238U 86019 CL 87055.93405650  .00009358  00000-0  00000-0 0 00369
2 17238 099.2951 138.4603 0071851 052.9538 307.8724 14.19426675014468
1 17305U 86019 CM 87055.97585904  .00004002  00000-0  00000-0 0 00146
2 17305 098.3851 119.7117 0316398 086.8268 276.9530 13.65037410014015
1 17306U 86019 CN 87055.95571063  .00010296  00000-0  00000-0 0 00115
2 17306 098.9301 136.6318 0056223 133.8284 226.8166 14.34157553014615
1 17307U 86019 CP 87055.97575580  .00002790  00000-0  00000-0 0 00149
2 17307 097.7207 114.8252 0237271 069.4720 293.2248 13.80476542014167
1 17308U 86019 CQ 87055.99199018 -.00000944  00000-0  00000-0 0 00162
2 17308 098.3961 119.3451 0317658 082.1952 281.5632 13.61559081013974
1 17309U 86019 CR 87055.97885377  .00002034  00000-0  00000-0 0 00106
2 17309 098.5839 121.2068 0399524 073.7917 290.7291 13.60505925013864
1 17310U 86019 CS 87055.96087362  .00007963  00000-0  00000-0 0 00145
2 17310 098.4174 117.3680 0472357 080.7069 284.7636 13.48363162013806
1 17311U 86019 CT 87055.95973249  .00000252  00000-0  00000-0 0 00103
2 17311 098.5683 131.8380 0012762 188.5452 171.6074 14.27998667014492
1 17312U 86019 CU 87055.98813192  .00001676  00000-0  00000-0 0 00108
2 17312 098.7120 132.4706 0031655 105.5933 254.9296 14.21787083014443
1 17313U 86019 CV 87055.98929406 -.00020944  00000-0  00000-0 0 00150
2 17313 098.5969 124.9667 0217504 087.7897 274.8670 13.85330374014010
1 17314U 86019 CW 87055.95979973  .00000788  00000-0  00000-0 0 00100
2 17314 098.8255 133.5152 0051748 011.1576 349.1286 14.19982897014412
1 17315U 86019 CX 87055.96314939  .00000966  00000-0  00000-0 0 00105
2 17315 098.6801 131.1984 0048871 076.9280 283.7893 14.16359492014300
1 17316U 86019 CY 87055.96333021  .00000364  00000-0  00000-0 0 00101
2 17316 098.5509 131.1521 0018519 134.3660 225.9598 14.25029088014461
1 17317U 86019 CZ 87055.93597792  .00088010  00000-0  00000-0 0 00223
2 17317 098.8472 144.3893 0203066 227.4288 131.0236 14.86799401015169
1 17318U 86019 DA 87055.95758140  .00061676  00000-0  00000-0 0 00175
2 17318 099.2696 147.2464 0155864 270.8451 087.5498 14.73390282014798
1 17319U 86019 DB 87055.97765101  .00005947  00000-0  00000-0 0 00105
2 17319 099.2624 143.7118 0129485 283.3932 075.3426 14.48747599014694
1 17320U 86019 DC 87055.93225327  .00003498  00000-0  00000-0 0 00147
2 17320 098.3528 131.7365 0108634 273.3398 085.5947 14.44389537014734
1 17321U 86019 DD 87055.95972383 -.00004163  00000-0  00000-0 0 00142
2 17321 099.1147 140.7073 0082205 270.6483 088.5854 14.40923394014648
1 17322U 86019 DE 87055.98131413  .00020715  00000-0  00000-0 0 00103
2 17322 098.6121 131.5170 0133586 154.7462 206.0888 14.26816680014602
1 17323U 86019 DF 87055.93375908  .00000414  00000-0  00000-0 0 00102
2 17323 099.0443 136.3095 0058117 003.4526 356.7591 14.21542095014415
1 17324U 86019 DG 87055.93892959 -.00000062  00000-0  00000-0 0 00103
2 17324 098.7504 131.7453 0055480 075.5030 285.2837 14.14722647014286
1 17336U 86019 DH 87055.95446567  .00001128  00000-0  00000-0 0 00088
2 17336 097.9791 122.3818 0097038 041.6753 319.2311 14.12831044014320
1 17337U 86019 DJ 87055.98088396  .00008213  00000-0  00000-0 0 00097
2 17337 098.6106 128.8483 0104795 083.0011 278.3606 14.08018907014382
1 17338U 86019 DK 87055.95356222  .00010879  00000-0  00000-0 0 00115
2 17338 097.8853 119.6859 0120036 044.1943 316.9282 14.03399100014231
1 17339U 86019 DL 87052.99597590  .00000147  00000-0  00000-0 0 00063
2 17339 098.1166 119.0468 0168101 107.9327 254.0782 13.97422983013753
1 17340U 86019 DM 87055.92992176  .00008321  00000-0  00000-0 0 00130
2 17340 099.8857 140.8480 0203340 072.3503 290.0266 13.95437500014221
1 17341U 86019 DN 87055.95312835  .00001883  00000-0  00000-0 0 00071
2 17341 097.8905 118.6407 0163885 083.4586 278.5739 13.94053395014916
1 17342U 86019 DP 87055.93487271  .00000789  00000-0  00000-0 0 00082
2 17342 098.2014 121.4320 0210566 114.0335 248.3648 13.89608237014302
1 17343U 86019 DQ 87054.94643981  .00004016  00000-0  00000-0 0 00093
2 17343 098.6433 124.4281 0290145 085.4129 278.0618 13.84643753014309
1 17344U 86019 DR 87055.95091406  .00000722  00000-0  00000-0 0 00084
2 17344 098.6419 132.3670 0043775 155.0062 205.3808 14.26239035014449
1 17345U 86019 DS 87055.96634212  .00001500  00000-0  00000-0 0 00090
2 17345 098.7805 135.1544 0037365 201.7290 158.2874 14.33966733014529
1 17346U 86019 DT 87055.94770085  .00001492  00000-0  00000-0 0 00087
2 17346 098.6543 132.7785 0020597 176.1760 184.0141 14.28625562014456
1 17347U 86019 DU 87055.95901722  .00037122  00000-0  00000-0 0 00149
2 17347 098.4946 135.1706 0119684 235.0655 123.9816 14.59016459014652
```

continued...

Appendix 1 (cont'd)

SPOT 1 R/B Cataloged Debris, February 25, 1987

```
1 17348U 86019 DV 87055.98227497  .00034183  00000-0  00000-0 0 00103
2 17348 099.1449 137.8205 0062163 322.1828 037.5553 14.31634889014595
1 17349U 86019 DW 87055.99551919 -.00058622  00000-0  00000-0 0 00176
2 17349 098.7172 131.5164 0141690 137.9701 223.3003 14.21942810014493
1 17350U 86019 DX 87055.97370451  .00000137  00000-0  00000-0 0 00072
2 17350 098.5911 129.3368 0088968 107.7239 253.4214 14.11268355014340
1 17351U 86019 DY 87055.96502165  .00002453  00000-0  00000-0 0 00085
2 17351 097.9046 120.5829 0120064 037.0692 323.9240 14.05704266014330
1 17352U 86019 DZ 87055.95050292  .00001065  00000-0  00000-0 0 00082
2 17352 097.8502 120.0814 0115910 034.6076 326.3116 14.07930227014215
1 17353U 86019 EA 87053.94854655  .00005089  00000-0  00000-0 0 00085
2 17353 097.9255 118.7789 0130275 053.5021 307.8608 14.04430868014389
1 17354U 86019 EB 87055.95768288  .00010847  00000-0  00000-0 0 00162
2 17354 097.8824 118.7591 0201503 043.2337 318.4944 13.98134731014104
1 17355U 86019 EC 87055.96100134  .00000232  00000-0  00000-0 0 00082
2 17355 098.5847 131.1551 0020096 103.9829 256.4141 14.22614410014441
1 17371U 86019 ED 87055.99294832  .00005454  00000-0  00000-0 0 00364
2 17371 099.0757 140.9553 0075629 255.3265 104.0105 14.46699667014725
1 17372U 86019 EE 87055.95730566  .00005731  00000-0  00000-0 0 00256
2 17372 098.5944 130.7482 0065483 141.2268 219.4178 14.21965766014280
1 17373U 86019 EF 87055.96086580  .00001385  00000-0  00000-0 0 00372
2 17373 098.8860 138.3829 0074891 232.3734 127.1215 14.44462048014741
1 17374U 86019 EG 87055.96734375  .00003889  00000-0  00000-0 0 00444
2 17374 097.9341 117.8254 0203129 092.0057 270.4903 13.84453641014037
1 17375U 86019 EH 87055.96004936  .00008788  00000-0  00000-0 0 00452
2 17375 097.8477 118.8637 0166225 045.1431 316.3639 14.00530975013853
1 17376U 86019 EJ 87055.99185894  .00004478  00000-0  00000-0 0 00596
2 17376 098.2562 122.7505 0181855 112.4875 249.6189 13.93952388013886
1 17377U 86019 EK 87055.93857699  .00011218  00000-0  00000-0 0 00220
2 17377 099.6210 147.0961 0217572 314.4025 044.0093 14.43886418014412
1 17378U 86019 EL 87055.99490618  .00006661  00000-0  00000-0 0 00294
2 17378 098.7159 134.2364 0028606 255.8230 104.0339 14.32936958014338
1 17379U 86019 EM 87055.96705105 -.00005941  00000-0  00000-0 0 00260
2 17379 099.2321 135.9428 0116504 045.3126 315.7999 14.06426688014132
1 17380U 86019 EN 87055.99989566  .00002954  00000-0  00000-0 0 00518
2 17380 097.9507 119.7558 0131995 080.3775 281.2814 13.97859323013866
1 17381U 86019 EP 87055.96752467 -.00139282  00000-0  00000-0 0 00419
2 17381 099.0227 143.1567 0150565 237.0350 121.6881 14.60003352014376
1 17382U 86019 EQ 87055.93902872  .00000600  00000-0  00000-0 0 00111
2 17382 098.5817 129.8083 0072163 116.9147 243.9974 14.15742124013867
1 17383U 86019 ER 87055.94241210  .00003662  00000-0  00000-0 0 00510
2 17383 098.2631 125.8538 0083436 026.9197 333.6820 14.15128944013876
1 17384U 86019 ES 87055.95863491  .00008070  00000-0  00000-0 0 00454
2 17384 098.7851 135.7708 0041850 244.3307 115.4075 14.37558374014257
1 17385U 86019 ET 87055.98904438  .00005546  00000-0  00000-0 0 00359
2 17385 098.6299 133.3994 0043783 276.7401 082.9368 14.33878675014501
1 17386U 86019 EU 87055.98318411  .00001170  00000-0  00000-0 0 00483
2 17386 097.9746 122.2201 0175872 007.7557 352.6821 14.12334911014102
1 17387U 86019 EV 87055.99510253  .00007330  00000-0  00000-0 0 00386
2 17387 099.1556 135.8684 0149800 021.3535 339.4340 14.12501453014130
1 17388U 86019 EW 87055.94735834  .00004193  00000-0  00000-0 0 00571
2 17388 098.9868 142.2005 0138819 236.3143 122.5338 14.59713569014215
1 17389U 86019 EX 87055.99277484  .00009350  00000-0  00000-0 0 00502
2 17389 099.0588 142.9472 0149600 230.3393 128.5113 14.59182329014236
1 17390U 86019 EY 87055.95747791  .00001691  00000-0  00000-0 0 00435
2 17390 098.2496 128.1277 0087080 326.5248 033.0999 14.29120641013884
1 17391U 86019 EZ 87055.98280723 -.00001203  00000-0  00000-0 0 00548
2 17391 098.3960 129.8495 0035285 340.6625 019.3777 14.27814055013921
1 17392U 86019 FA 87055.99845970  .00000989  00000-0  00000-0 0 00203
2 17392 099.1114 137.4930 0042753 001.0194 359.1617 14.23621927014039
1 17393U 86019 FB 87055.97341256 -.00037219  00000-0  00000-0 0 00555
2 17393 099.4721 136.3124 0221394 067.6545 294.8435 13.91204355013564
1 17394U 86019 FC 87055.93439017  .00004928  00000-0  00000-0 0 00500
2 17394 099.0162 141.4230 0109222 235.1159 124.0308 14.53206187014606
1 17395U 86019 FD 87055.94304424  .00006005  00000-0  00000-0 0 00569
2 17395 098.9987 140.4168 0110650 231.3386 127.8441 14.48843130014581
1 17396U 86019 FE 87055.95749408  .00006687  00000-0  00000-0 0 00281
2 17396 098.8446 136.7432 0043314 245.1260 114.5972 14.37316936014421
1 17397U 86019 FF 87054.94502812  .00007270  00000-0  00000-0 0 00364
2 17397 099.0296 135.9101 0100746 339.9167 019.8617 14.27202032014268
1 17421U 86019 FG 87055.97048113  .00005080  00000-0  00000-0 0 00070
2 17421 098.8722 138.1926 0080548 238.3894 120.9990 14.44626511014587
1 17422U 86019 FH 87055.98605427  .00012470  00000-0  00000-0 0 00081
2 17422 098.7984 134.2681 0048660 150.9398 209.5067 14.30463710014437
1 17423U 86019 FJ 87055.96149584 -.00000021  00000-0  00000-0 0 00065
2 17423 099.2215 132.0414 0205300 076.4654 285.9801 13.84266342013910
```

continued...

Appendix 1 (cont'd)

SPOT 1 R/B Cataloged Debris, February 25, 1987

```
1 17424U 86019 FK 87055.95317407  .00008718  00000-0  00000-0 0 00088
2 17424 098.5947 131.3704 0055431 153.4619 206.9976 14.25868647014328
1 17425U 86019 FL 87055.96149401  .00000040  00000-0  00000-0 0 00065
2 17425 099.2201 132.0432 0205089 076.4019 286.0406 13.84267451013911
1 17426U 86019 FM 87055.97413442  .00000376  00000-0  00000-0 0 00068
2 17426 098.5728 131.3694 0038615 145.9208 214.5020 14.24885460014326
1 17427U 86019 FN 87055.96294012  .00001495  00000-0  00000-0 0 00072
2 17427 098.2396 127.3434 0023666 008.9165 351.2990 14.24590542014398
1 17428U 86019 FP 87055.99395134 -.00006896  00000-0  00000-0 0 00119
2 17428 098.6449 131.7640 0042794 346.8501 013.2115 14.22326515014233
1 17429U 86019 FQ 87055.94317519 -.00000016  00000-0  00000-0 0 00068
2 17429 099.0759 133.4027 0119590 064.2663 297.1325 14.01993040014292
1 17430U 86019 FR 87055.93204964  .00010286  00000-0  00000-0 0 00089
2 17430 099.1346 144.2846 0153525 236.2357 122.4721 14.61611628014399
1 17431U 86019 FS 87055.93196666  .00006330  00000-0  00000-0 0 00062
2 17431 098.6649 133.1028 0013756 235.5906 124.4536 14.29968044014300
1 17432U 86019 FT 87055.99302173  .00022502  00000-0  00000-0 0 00093
2 17432 099.0333 145.5914 0205597 217.2737 141.4648 14.75276662014641
1 17433U 86019 FU 87055.95517829  .00001491  00000-0  00000-0 0 00067
2 17433 098.9015 137.5994 0059285 259.6575 099.8546 14.38683517014651
1 17434U 86019 FV 87055.95347457 -.00000549  00000-0  00000-0 0 00060
2 17434 098.8151 135.7563 0105541 314.5761 044.7408 14.33698819014112
1 17435U 86019 FW 87055.98547030  .00005770  00000-0  00000-0 0 00063
2 17435 097.9819 122.7299 0096958 016.7828 343.7069 14.14241918014312
1 17436U 86019 FX 87055.97724013  .00001286  00000-0  00000-0 0 00070
2 17436 098.6572 132.4306 0022721 132.6787 227.6869 14.25993130013974
1 17437U 86019 FY 87055.95071442  .00008334  00000-0  00000-0 0 00071
2 17437 098.6099 132.7573 0036024 238.1199 121.7044 14.34329625013913
1 17438U 86019 FZ 87055.97478387  .00001388  00000-0  00000-0 0 00075
2 17438 097.9826 118.6980 0199304 090.9703 271.4825 13.87013368013466
1 17439U 86019 GA 87055.95966560  .00002965  00000-0  00000-0 0 00071
2 17439 098.8935 140.2714 0122846 231.6417 127.4270 14.54559178014483
1 17440U 86019 GB 87055.96016712  .00008794  00000-0  00000-0 0 00075
2 17440 099.3103 145.3021 0142456 304.2061 054.6276 14.57561952014478
1 17441U 86019 GC 87055.94677043  .00000252  00000-0  00000-0 0 00569
2 17441 099.4414 142.4707 0201717 338.4070 020.9245 14.29496380014304
1 17442U 86019 GD 87054.98116891  .00004676  00000-0  00000-0 0 00432
2 17442 098.3263 126.3111 0040277 087.1005 273.5334 14.18789410013674
1 17443U 86019 GE 87055.96202092  .00000611  00000-0  00000-0 0 00379
2 17443 098.7755 131.3955 0191129 146.5916 214.8034 14.11520302014123
1 17444U 86019 GF 87055.98619751  .00000410  00000-0  00000-0 0 00362
2 17444 097.9426 118.1189 0207779 073.6822 288.7604 13.86150333013459
1 17445U 86019 GG 87055.98737767  .00008209  00000-0  00000-0 0 00499
2 17445 098.4266 125.6073 0171586 126.1110 235.6619 14.01161406013664
1 17446U 86019 GH 87055.98659046  .00004810  00000-0  00000-0 0 00475
2 17446 098.7744 134.5105 0018965 269.4296 090.5274 14.29941770014375
1 17447U 86019 GJ 87055.97537328  .00008588  00000-0  00000-0 0 00514
2 17447 099.0960 144.8278 0171171 229.4382 129.2401 14.66247182013778
1 17448U 86019 GK 87055.94047943  .00009403  00000-0  00000-0 0 00530
2 17448 098.9565 143.1922 0177755 219.6665 139.2013 14.66760586013863
1 17449U 86019 GL 87055.98831089  .00003113  00000-0  00000-0 0 00269
2 17449 098.5346 130.8873 0039888 141.7244 218.7337 14.24926632013702
1 17450U 86019 GM 87055.95999358  .00002527  00000-0  00000-0 0 00394
2 17450 097.8335 118.6738 0147253 030.4819 330.5347 13.99993867013435
1 17451U 86019 GN 87055.97230046  .00009041  00000-0  00000-0 0 00498
2 17451 098.0828 119.6362 0210597 056.5154 305.6466 13.87861569013337
1 17452U 86019 GP 87049.21260665  .00000350  00000-0  00000-0 0 00088
2 17452 098.8226 129.4519 0049960 277.0499 082.5574 14.37749714013735
1 17453U 86019 GQ 87055.97578976  .00003579  00000-0  00000-0 0 00172
2 17453 099.1115 138.6185 0114096 336.3652 023.2870 14.30897629013442
1 17454U 86019 GR 87055.94261070  .00003878  00000-0  00000-0 0 00345
2 17454 098.6752 132.9294 0025891 195.7355 164.3584 14.28420162013780
1 17455U 86019 GS 87055.98536099  .00000400  00000-0  00000-0 0 00322
2 17455 098.0158 120.0409 0168330 065.6954 296.2219 13.93734286013592
1 17456U 86019 GT 87055.99045810 -.00032244  00000-0  00000-0 0 00287
2 17456 098.8198 136.7472 0052323 234.3505 125.3293 14.41092605013251
1 17457U 86019 GU 87055.96243439  .00007694  00000-0  00000-0 0 00372
2 17457 098.5229 128.3448 0091495 112.3340 248.8107 14.10970244013199
1 17458U 86019 GV 87055.94983312  .00008189  00000-0  00000-0 0 00372
2 17458 097.8823 117.8878 0166229 091.6746 270.3995 13.90681880013501
1 17459U 86019 GW 87055.99743142  .00000621  00000-0  00000-0 0 00219
2 17459 098.6059 131.2492 0045671 128.7357 231.8472 14.21648406013538
1 17460U 86019 GX 87055.97932517  .00006524  00000-0  00000-0 0 00369
2 17460 098.3479 125.9080 0167949 017.8370 342.9143 14.07884712013177
1 17461U 86019 GY 87055.96455953  .00001077  00000-0  00000-0 0 00192
2 17461 098.6333 132.5924 0019793 173.3641 186.8365 14.28009533013076
```

continued...

Appendix 1 (cont'd)

SPOT 1 R/B Cataloged Debris, February 25, 1987

```
1 17462U 86019 GZ 87055.93977311  .00009837  00000-0  00000-0 0 00335
2 17462 098.6859 132.2625 0079389 148.3699 212.2842 14.24458138013380
1 17463U 86019 HA 87055.97840159  .00005304  00000-0  00000-0 0 00389
2 17463 099.0232 141.8631 0139879 232.6517 126.2451 14.55135487014633
1 17464U 86019 HB 87055.96583189  .00002506  00000-0  00000-0 0 00411
2 17464 098.6906 129.0569 0097497 082.0891 279.1868 14.04035764012792
1 17465U 86019 HC 87055.99938333  .00006339  00000-0  00000-0 0 00441
2 17465 098.1394 121.6735 0173115 110.1175 251.9246 13.96194609012783
1 17466U 86019 HD 87053.99292698 -.00394820  00000-0  00000-0 0 00230
2 17466 098.4210 128.3974 0038309 186.1309 173.9972 14.28479115012952
1 17467U 86019 HE 87055.94103393  .00014587  00000-0  00000-0 0 00330
2 17467 098.5883 131.1844 0039108 175.3043 184.9069 14.25077741013182
1 17468U 86019 HF 87055.95133095  .00006845  00000-0  00000-0 0 00253
2 17468 098.7677 135.4105 0051304 228.5764 131.1574 14.36795222012576
1 17469U 86019 HG 87055.95585915 -.00004662  00000-0  00000-0 0 00353
2 17469 098.7593 130.6396 0252126 011.0327 349.6759 14.07953334012386
1 17470U 86019 HH 87055.94632402  .00000780  00000-0  00000-0 0 00275
2 17470 098.3651 130.5327 0097313 316.8202 042.5950 14.36391159012560
1 17471U 86019 HJ 87055.93379685  .00001591  00000-0  00000-0 0 00176
2 17471 098.7234 130.7097 0073860 084.2219 276.7911 14.11039496012994
1 17472U 86019 HK 87055.93761615  .00000268  00000-0  00000-0 0 00221
2 17472 097.9982 118.8075 0200749 089.5017 272.9668 13.86592850012223
1 17473U 86019 HL 87055.95579513  .00002228  00000-0  00000-0 0 00169
2 17473 098.5684 130.7395 0037572 136.1835 224.2920 14.22363608014284
1 17474U 86019 HM 87055.96195498  .00007717  00000-0  00000-0 0 00581
2 17474 098.7073 134.5424 0009386 174.8154 185.3693 14.36797872013584
1 17475U 86019 HN 87055.98024702  .00007940  00000-0  00000-0 0 00402
2 17475 098.6514 133.6926 0098438 171.2504 189.0990 14.35475307012414
1 17476U 86019 HP 87055.97829889  .00003571  00000-0  00000-0 0 00665
2 17476 098.5877 129.5649 0079784 097.5192 263.5601 14.13477697012662
1 17477U 86019 HQ 87055.94721536  .00005680  00000-0  00000-0 0 00172
2 17477 098.5765 131.2086 0071554 146.6126 214.0153 14.24987789012678
1 17478U 86019 HR 87055.95235488  .00001274  00000-0  00000-0 0 00197
2 17478 098.7181 131.2279 0050844 083.8266 276.9246 14.16444300013098
1 17479U 86019 HS 87035.62546868  .00032439  00000-0  00000-0 0 00015
2 17479 098.6494 111.1478 0085492 175.3289 184.9260 14.16681392009668
1 17539U 86019 HT 87055.97137723 -.00002237  00000-0  00000-0 0 00023
2 17539 098.4882 130.5036 0024889 335.3630 024.6918 14.25629119014317
1 17540U 86019 HU 87055.96530451  .00002595  00000-0  00000-0 0 00025
2 17540 097.9457 119.5437 0167446 097.8066 264.2680 13.96000678014052
1 17541U 86019 HV 87055.95120080  .00005462  00000-0  00000-0 0 00024
2 17541 099.0538 142.6286 0156283 226.4121 132.4599 14.57106884014334
1 17542U 86019 HW 87055.96169959  .00007412  00000-0  00000-0 0 00021
2 17542 098.7046 133.3332 0029840 209.7526 150.2522 14.29653958013910
1 17543U 86019 HX 87055.95281276 -.00000123  00000-0  00000-0 0 00021
2 17543 098.9406 131.1742 0133195 081.3613 280.3146 13.97979520013775
1 17544U 86019 HY 87055.96003389  .00012284  00000-0  00000-0 0 00020
2 17544 099.2610 133.1729 0147901 100.0125 261.8289 13.92176540013836
1 17545U 86019 HZ 87055.98358934  .00001547  00000-0  00000-0 0 00022
2 17545 098.4503 129.8504 0040694 356.7160 003.4305 14.24563987013618
1 17546U 86019 JA 87054.94047194  .00000465  00000-0  00000-0 0 00018
2 17546 098.7981 134.3025 0028342 280.0848 079.7700 14.32607751013688
1 17547U 86019 JB 87055.93485018  .00002079  00000-0  00000-0 0 00024
2 17547 097.9197 120.0660 0137063 046.3774 314.9208 14.01905314012901
1 17548U 86019 JC 87055.97405904  .00000524  00000-0  00000-0 0 00028
2 17548 098.6337 131.6219 0031179 137.5668 222.8485 14.23029995013114
1 17549U 86019 JD 87055.94841181  .00003382  00000-0  00000-0 0 00022
2 17549 098.7636 135.0858 0032753 245.9380 113.8940 14.34220634014660
1 17550U 86019 JE 87055.95888695  .00008201  00000-0  00000-0 0 00021
2 17550 098.6784 134.0376 0056062 246.1830 113.3985 14.35340957012250
1 17551U 86019 JF 87055.98158443  .00000494  00000-0  00000-0 0 00022
2 17551 098.6057 131.7644 0016669 124.2732 236.0586 14.24813832012268
1 17552U 86019 JG 87055.98288554  .00005530  00000-0  00000-0 0 00015
2 17552 098.5899 131.3643 0056502 133.4946 227.1507 14.24414423011969
1 17553U 86019 JH 87055.95298998  .00006355  00000-0  00000-0 0 00023
2 17553 098.1207 119.9306 0202320 091.1819 271.3052 13.86169392011490
1 17554U 86019 JJ 87055.93256959  .00001680  00000-0  00000-0 0 00029
2 17554 098.6629 131.4737 0059520 120.5098 240.2524 14.19575959011574
1 17555U 86019 JK 87055.95859435  .00001296  00000-0  00000-0 0 00023
2 17555 097.9686 117.7266 0225746 092.4100 270.3443 13.81344879010936
1 17556U 86019 JL 87055.93997778  .00002945  00000-0  00000-0 0 00027
2 17556 098.5437 129.7018 0077169 134.2547 226.5548 14.17601513011335
1 17557U 86019 JM 87054.97577789  .00001551  00000-0  00000-0 0 00018
2 17557 098.8786 136.1670 0068908 286.3859 073.0327 14.37946433014156
1 17558U 86019 JN 87055.97865042  .00016418  00000-0  00000-0 0 00021
2 17558 099.0710 143.3685 0158732 218.5565 140.4795 14.60941646010970
```

Appendix 2
SPOT 1 R/B Uncataloged Debris, February 25, 1987

```
1 87086U 86019     87055.96933012  .00007732  00000-0  00000-0 0 00025
2 87086 098.5524 130.9257 0085011 112.1717 248.9064 14.25695879000881
1 87132U 86019     87055.93081455  .00005393  00000-0  00000-0 0 00019
2 87132 098.4306 129.3734 0030549 129.8426 230.6004 14.23880731000793
1 87133U 86019     87055.94397767 -.00033651  00000-0  00000-0 0 00038
2 87133 098.5801 130.7042 0113453 147.0133 213.8746 14.24785547000501
1 87134U 86019     87055.98440514  .00003846  00000-0  00000-0 0 00012
2 87134 098.7551 121.4805 0388873 093.5924 271.0256 13.50802539000706
1 87622U 86019     87055.98460454  .00000268  00000-0  00000-0 0 00069
2 87622 098.0267 118.3029 0218189 081.8293 280.8070 13.80853233005278
1 87739U 86019     87055.98032881 -.00019405  00000-0  00000-0 0 00128
2 87739 098.6133 131.8452 0082016 170.4558 189.8764 14.26862701007868
1 87740U 86019     87055.99203671  .00000062  00000-0  00000-0 0 00036
2 87740 098.6493 131.2873 0037095 087.7005 272.8970 14.18801564007957
1 87742U 86019     87042.94204440  .00003413  00000-0  00000-0 0 00015
2 87742 098.5974 118.7112 0041566 175.9532 184.2550 14.24771768005868
1 88000U 86019     87055.94876544  .00001820  00000-0  00000-0 0 00084
2 88000 097.8795 117.7644 0184147 100.7935 261.4544 13.89462451014186
1 88001U 86019     87055.99291946  .00007031  00000-0  00000-0 0 00201
2 88001 099.2801 144.8724 0176233 285.2270 073.0079 14.54713156014330
1 88002U 86019     87053.96951783  .00004192  00000-0  00000-0 0 00079
2 88002 098.4188 123.5002 0137710 100.9500 260.7730 13.98980683010792
1 88003U 86019     87055.97712577  .00002366  00000-0  00000-0 0 00160
2 88003 098.6127 131.2835 0068942 135.3657 225.3655 14.21812415014288
1 88004U 86019     87055.96236745  .00004867  00000-0  00000-0 0 00144
2 88004 098.7216 135.4242 0060951 247.8943 111.6336 14.40184751011024
1 88005U 86019     87052.99030377 -.00001459  00000-0  00000-0 0 00008
2 88005 098.7811 129.2311 0058187 088.3141 272.5245 14.15397693010354
1 88006U 86019     87055.93195140 -.00001586  00000-0  00000-0 0 00063
2 88006 098.4314 125.9538 0154159 120.5927 241.1088 14.01930400010183
1 88007U 86019     87032.94599121  .00004299  00000-0  00000-0 0 00026
2 88007 098.0016 100.4727 0167430 164.5838 196.1069 13.98993356006875
1 88008U 86019     87015.97263580  .00007872  00000-0  00000-0 0 00028
2 88008 098.6840 094.7849 0066357 002.4357 357.7707 14.43837084005137
1 88009U 86019     87055.94474347  .00002243  00000-0  00000-0 0 00190
2 88009 098.5472 130.8903 0052685 139.7748 220.7905 14.24555843012178
1 88010U 86019     87055.93385360 -.00002179  00000-0  00000-0 0 00141
2 88010 097.8275 115.1842 0233231 061.5383 300.9573 13.77097327010890
1 88011U 86019     87054.95774000  .00030379  00000-0  00000-0 0 00171
2 88011 098.6887 131.4622 0026094 120.1519 240.2778 14.24422443012244
1 88012U 86019     87055.96527039  .00003757  00000-0  00000-0 0 00105
2 88012 098.1832 124.4077 0091000 093.2887 267.9247 14.10299395009542
1 88013U 86019     87055.94867461  .00084737  00000-0  00000-0 0 00051
2 88013 098.7956 141.3006 0118893 257.3083 101.5399 14.72785053010178
1 88014U 86019     87055.95154276  .00002066  00000-0  00000-0 0 00174
2 88014 098.6269 131.9467 0019470 071.5654 288.8196 14.25242915008637
1 88015U 86019     87055.97013283  .00003806  00000-0  00000-0 0 00082
2 88015 098.7040 131.1200 0061955 082.4027 278.4725 14.14386047009427
1 88016U 86019     87053.94163798  .00019288  00000-0  00000-0 0 00019
2 88016 099.1298 143.5223 0178336 233.0359 125.4993 14.70213251008250
1 88017U 86019     87054.94822438  .00000420  00000-0  00000-0 0 00090
2 88017 098.1054 119.5195 0193198 075.5814 286.7229 13.89472033008414
1 88018U 86019     87055.99289354  .00000144  00000-0  00000-0 0 00128
2 88018 098.6506 132.5671 0044251 155.4086 204.9776 14.26613464008359
1 88019U 86019     87055.96682205  .00002287  00000-0  00000-0 0 00106
2 88019 098.5661 131.2151 0049129 136.5466 224.0159 14.25078094008767
1 88020U 86019     87043.93289184 -.00003894  00000-0  00000-0 0 00017
2 88020 098.7434 119.5636 0031723 183.5954 176.5552 14.17501525005172
1 88021U 86019     87055.98903017 -.00001050  00000-0  00000-0 0 00173
2 88021 097.8352 118.5367 0133581 051.5714 309.7896 13.98346371009736
1 88034U 86019     87055.97870508  .00005706  00000-0  00000-0 0 00029
2 88034 098.0293 119.4049 0180348 096.2441 265.9827 13.89382748001287
1 88054U 86019     87054.98328782  .00015730  00000-0  00000-0 0 00122
2 88054 098.9036 137.5685 0074359 238.0086 121.4434 14.44720613007021
1 88056U 86019     87055.93689334  .00022344  00000-0  00000-0 0 00076
2 88056 098.7214 133.9714 0056806 179.9667 180.2094 14.35129195007470
1 88065U 86019     87055.98547667  .00000787  00000-0  00000-0 0 00081
2 88065 098.5311 131.0035 0034805 148.2078 212.1772 14.25781560007210
1 88101U 86019     87049.96821465  .00001944  00000-0  00000-0 0 00068
2 88101 098.5284 124.9060 0053540 147.6901 212.8134 14.24670378005292
1 88115U 86019     87042.99629271 -.00002393  00000-0  00000-0 0 00026
2 88115 099.0722 128.1070 0116187 276.9555 081.9012 14.52561903003848
```

continued...

Appendix 2 (cont'd)

SPOT 1 R/B Uncataloged Debris, February 25, 1987

```
1 88116U 86019     87052.95630294  .00007127  00000-0  00000-0 0 00020
2 88116 099.5690 139.4278 0094592 037.2624 323.5594 14.23569066004929
1 88579U 86019     87055.93719397  .00007541  00000-0  00000-0 0 00138
2 88579 098.5558 131.2743 0045769 126.6734 233.9223 14.26546648009942
1 88614U 86019     87055.95363473  .00021133  00000-0  00000-0 0 00199
2 88614 098.7343 136.3281 0128888 278.5984 080.1200 14.48233478008783
1 88615U 86019     87055.93647207  .00035686  00000-0  00000-0 0 00152
2 88615 097.7291 118.4947 0158646 041.7649 319.6047 14.07683017008645
1 88621U 86019     87054.95915117  .00001546  00000-0  00000-0 0 00101
2 88621 098.7147 133.7674 0053138 123.0641 237.6176 14.38678503008286
1 88622U 86019     87055.99596286 -.00000382  00000-0  00000-0 0 00118
2 88622 098.7791 135.6329 0050803 247.4645 112.1726 14.36167111008274
1 88623U 86019     87042.97315481  .00000000  00000-0  00000-0 0 00003
2 88623 098.2405 119.4014 0076551 052.0467 308.8154 14.27073407006870
1 88672U 86019     87055.98557672  .00006931  00000-0  00000-0 0 00225
2 88672 098.6909 132.2881 0060503 121.8205 238.9439 14.24571039014359
1 88674U 86019     87043.93344025  .00069963  00000-0  00000-0 0 00049
2 88674 098.7246 119.2517 0099919 177.0697 183.1645 14.21941381012624
1 88677U 86019     87055.94317451  .00008195  00000-0  00000-0 0 00241
2 88677 098.7602 132.6091 0132713 136.3054 224.9274 14.25002688014316
1 88682U 86019     87055.99866686  .00012656  00000-0  00000-0 0 00260
2 88682 099.1386 145.8268 0186453 227.6132 130.9753 14.68865451014870
1 88683U 86019     87055.95626109  .00030707  00000-0  00000-0 0 00347
2 88683 099.6933 151.5082 0191816 280.7653 077.2609 14.65022897014946
1 88689U 86019     87055.93261673  .00004659  00000-0  00000-0 0 00181
2 88689 097.9732 116.7083 0310166 055.9225 307.1470 13.76253943013851
1 88690U 86019     87055.93574969  .00200857  00000-0  00000-0 0 00298
2 88690 099.0301 147.5514 0107854 230.0359 129.1944 14.96484266014958
1 88691U 86019     87055.97509410  .00002447  00000-0  00000-0 0 00264
2 88691 099.2069 146.4079 0177795 221.5955 137.2201 14.71823082014585
1 88693U 86019     87055.95751160  .00009030  00000-0  00000-0 0 00271
2 88693 098.8876 130.7258 0161465 109.8210 252.0928 13.99811407013849
1 88695U 86019     87055.98010594  .00005777  00000-0  00000-0 0 00325
2 88695 098.7889 127.3774 0164524 100.2419 261.7864 13.86913612013836
1 88698U 86019     87055.93197063  .00004940  00000-0  00000-0 0 00129
2 88698 098.9489 135.6495 0023413 069.0675 291.3560 14.25103574014246
1 88699U 86019     87055.99834754  .00000061  00000-0  00000-0 0 00110
2 88699 098.9131 130.7592 0137843 080.8755 280.8507 13.97305502013840
1 88701U 86019     87055.97211915  .00088682  00000-0  00000-0 0 00238
2 88701 098.9764 144.9978 0167761 229.7038 129.0005 14.81801006014911
1 88702U 86019     87055.94060941  .00023194  00000-0  00000-0 0 00201
2 88702 099.3072 147.2711 0170841 240.7331 117.7287 14.66881102014152
1 88703U 86019     87055.96077736  .00009854  00000-0  00000-0 0 00155
2 88703 098.6194 135.4382 0092005 245.3321 113.8846 14.47238323014639
1 88705U 86019     87055.98312070  .00003719  00000-0  00000-0 0 00134
2 88705 098.8433 136.4782 0037579 251.5483 108.2181 14.36571933014250
1 88706U 86019     87055.95643692  .00004651  00000-0  00000-0 0 00269
2 88706 098.8046 135.5864 0053750 208.4576 151.4255 14.34777894014383
1 88708U 86019     87055.99852952  .00005119  00000-0  00000-0 0 00143
2 88708 098.4686 131.5150 0026162 186.6938 173.4466 14.34005424014348
1 88710U 86019     87055.98214477  .00003700  00000-0  00000-0 0 00109
2 88710 098.5642 132.1046 0150169 321.2821 037.8238 14.30676643014600
1 88711U 86019     87055.99632176  .00000389  00000-0  00000-0 0 00145
2 88711 098.3400 129.1688 0018681 355.8565 004.3022 14.28754247014580
1 88712U 86019     87055.99776368  .00000802  00000-0  00000-0 0 00205
2 88712 098.2723 127.9519 0036381 341.0792 018.9596 14.26607363014326
1 88714U 86019     87055.96477410  .00003347  00000-0  00000-0 0 00173
2 88714 098.7056 130.2888 0115021 103.0728 258.3853 14.10006422013969
1 88718U 86019     87055.93514264  .00002992  00000-0  00000-0 0 00134
2 88718 098.8432 136.7396 0051181 252.0683 107.5486 14.38309116014298
1 88719U 86019     87055.94405890 -.00043626  00000-0  00000-0 0 00398
2 88719 099.8718 147.2287 0115634 354.7003 005.3495 14.32544802014082
1 88723U 86019     87055.98521646  .00039234  00000-0  00000-0 0 00181
2 88723 099.2993 147.6110 0151910 240.3098 118.3498 14.70145658014475
1 88724U 86019     87055.99045505  .00002953  00000-0  00000-0 0 00164
2 88724 099.0248 142.0267 0139415 224.9614 134.0808 14.55934443014647
1 88725U 86019     87055.95346091  .00020026  00000-0  00000-0 0 00324
2 88725 098.6539 133.7534 0061739 184.2196 175.9044 14.38550297014071
1 88729U 86019     87054.97416437  .00007339  00000-0  00000-0 0 00480
2 88729 098.7041 130.9820 0108733 107.8911 253.4707 14.22375650013925
1 88733U 86019     87055.95550617  .00000952  00000-0  00000-0 0 00128
2 88733 097.9902 118.9574 0193284 091.4492 270.9364 13.88213006013959
1 88742U 86019     87055.99176602  .00004061  00000-0  00000-0 0 00462
2 88742 097.8932 119.7687 0124865 051.9814 309.3092 14.02970486013676
1 88743U 86019     87055.98608939  .00007615  00000-0  00000-0 0 00362
2 88743 098.9255 131.2813 0167968 114.3539 247.5798 14.00262029013576
```

continued...

Appendix 2 (cont'd)
SPOT 1 R/B Uncataloged Debris, February 25, 1987

1	88747U	86019	87055.97212208	.00068538	00000-0	00000-0	0	00265
2	88747	099.5886	150.5706	0125570	253.6967	105.0977	14.70962726014797	
1	88748U	86019	87055.97901147	.00033281	00000-0	00000-0	0	00271
2	88748	098.5102	136.2561	0134958	248.9067	109.8248	14.61302015014428	
1	88750U	86019	87055.96683551	.00004293	00000-0	00000-0	0	00157
2	88750	099.1900	143.3537	0121329	266.7133	092.0751	14.51724429014446	
1	88752U	86019	87055.95644405	-.00029941	00000-0	00000-0	0	00522
2	88752	098.7186	135.8385	0078992	224.4716	135.0717	14.43039702014344	
1	88753U	86019	87055.98319533	.00020572	00000-0	00000-0	0	00392
2	88753	098.8623	137.5188	0056123	244.4436	115.1516	14.45135394013953	
1	88754U	86019	87055.94535847	.00000069	00000-0	00000-0	0	00194
2	88754	098.7072	135.2671	0056148	259.8424	099.6995	14.39043864014464	
1	88755U	86019	87055.97165828	.00003372	00000-0	00000-0	0	00130
2	88755	098.8207	136.8137	0053945	254.3611	105.2188	14.39875544014332	
1	88756U	86019	87055.98381871	.00001963	00000-0	00000-0	0	00156
2	88756	098.3743	130.8271	0036572	258.8498	100.9144	14.36526688014192	
1	88758U	86019	87055.98587819	-.00001973	00000-0	00000-0	0	00464
2	88758	098.6525	133.3256	0095555	171.0989	189.2479	14.31422042013827	
1	88760U	86019	87055.98578227	.00001465	00000-0	00000-0	0	00230
2	88760	098.6465	132.4607	0021019	160.8397	199.4137	14.26761125013989	
1	88767U	86019	87055.94268117	.00243077	00000-0	00000-0	0	00380
2	88767	099.2555	150.0337	0175407	245.8192	112.5205	14.97571429014591	
1	88768U	86019	87055.96292619	.00029855	00000-0	00000-0	0	00293
2	88768	099.2266	147.8079	0152094	227.6159	131.2682	14.76122511014189	
1	88769U	86019	87055.99623467	.00084547	00000-0	00000-0	0	00295
2	88769	098.4642	137.1203	0171591	204.5177	154.8397	14.75602068014090	
1	88770U	86019	87055.95649065	.00061850	00000-0	00000-0	0	00691
2	88770	099.6060	149.5271	0184978	231.4294	127.0814	14.61188507014158	
1	88774U	86019	87055.95303884	.00003653	00000-0	00000-0	0	00216
2	88774	098.7996	137.9211	0100339	237.6383	121.5643	14.48518114014639	
1	88775U	86019	87055.95227096	-.00003240	00000-0	00000-0	0	00445
2	88775	098.9400	138.8682	0090334	223.4030	136.0593	14.43427691014199	
1	88776U	86019	87055.96974266	.00003422	00000-0	00000-0	0	00132
2	88776	098.7978	135.7039	0046375	246.7940	112.8924	14.35841789013934	
1	88777U	86019	87055.98375594	.00008566	00000-0	00000-0	0	00486
2	88777	098.6270	133.2878	0050393	203.9298	156.0110	14.34308203014272	
1	88778U	86019	87055.98050031	-.00001162	00000-0	00000-0	0	00423
2	88778	098.4668	131.1493	0072237	313.9169	045.6631	14.31577908013820	
1	88780U	86019	87055.98259286	.00002482	00000-0	00000-0	0	00569
2	88780	098.1662	125.7162	0071644	005.2525	354.9945	14.20106803014199	
1	88781U	86019	87055.95410751	.00028107	00000-0	00000-0	0	00206
2	88781	099.3224	148.8090	0210440	240.3614	117.7081	14.74838059014160	
1	88782U	86019	87055.99996587	.00014846	00000-0	00000-0	0	00189
2	88782	099.3047	148.6580	0183207	242.9953	115.3038	14.73546567013861	
1	88783U	86019	87055.97173735	.00084700	00000-0	00000-0	0	00243
2	88783	099.4247	150.6639	0158920	242.3612	116.1978	14.79452748014173	
1	88784U	86019	87066.99634643	.00065154	00000-0	00000-0	0	00162
2	88784	098.8092	142.4023	0145965	233.6366	125.1893	14.76099223014173	
1	88787U	86019	87055.98656512	.00003969	00000-0	00000-0	0	00155
2	88787	099.0531	141.9567	0110795	237.6121	121.4893	14.53355395013730	
1	88788U	86019	87055.98427290	.00003546	00000-0	00000-0	0	00145
2	88788	098.9858	138.8913	0063293	260.3756	099.0845	14.40607082013464	
1	88789U	86019	87055.95459976	.00009332	00000-0	00000-0	0	00340
2	88789	098.3700	129.6703	0030937	287.0292	072.8069	14.33245737013767	
1	88790U	86019	87055.98554492	-.00000593	00000-0	00000-0	0	00233
2	88790	098.6071	132.1092	0031307	182.9420	177.2143	14.28326444013651	
1	88795U	86019	87055.95286364	.00088426	00000-0	00000-0	0	00235
2	88795	098.8092	143.1379	0203511	232.7096	125.6012	14.82859210013847	
1	88796U	86019	87055.93459352	-.00002671	00000-0	00000-0	0	00209
2	88796	099.0295	140.8081	0114659	235.6078	123.4804	14.48369296013685	
1	88797U	86019	87055.95508478	.00003188	00000-0	00000-0	0	00142
2	88797	098.9355	139.0546	0070406	238.0692	121.4209	14.45175548014384	
1	88798U	86019	87055.98537868	.00003785	00000-0	00000-0	0	00111
2	88798	098.7010	135.8758	0063816	282.2720	077.1900	14.43836312013802	
1	88799U	86019	87055.95254870	.00007262	00000-0	00000-0	0	00190
2	88799	098.7372	135.3443	0058370	227.0273	132.6579	14.38518862013584	
1	88801U	86019	87055.99727118	.00016030	00000-0	00000-0	0	00337
2	88801	098.8925	136.3980	0037165	296.6211	063.1733	14.35720800013992	
1	88804U	86019	87055.98388023	-.00004668	00000-0	00000-0	0	00310
2	88804	098.8074	134.1169	0059783	154.2518	206.2218	14.27710041014181	
1	88805U	86019	87055.96418418	.00003999	00000-0	00000-0	0	00147
2	88805	099.3554	138.4910	0077800	045.9802	314.8288	14.13685819013703	

continued...

Appendix 2 (cont'd)

SPOT 1 R/B Uncataloged Debris, February 25, 1987

```
1 88806U 86019    87055.93405323  .00004020  00000-0  00000-0 0 00153
2 88806 098.5647 129.1747 0073287 099.5523 261.4488 14.13156730013303
1 88807U 86019    87055.98504783 -.00002260  00000-0  00000-0 0 00435
2 88807 097.8982 118.7063 0154428 082.1120 279.8075 13.93638413013597
1 88809U 86019    87055.96846843  .00007296  00000-0  00000-0 0 00142
2 88809 098.6380 132.3191 0036003 163.1889 197.1053 14.27553200013815
1 88810U 86019    87055.95905526 -.00010707  00000-0  00000-0 0 00240
2 88810 099.1049 135.4371 0068142 007.8042 352.4721 14.14753606013730
1 88811U 86019    87055.96957621 -.00015024  00000-0  00000-0 0 00231
2 88811 097.9942 122.2385 0053396 036.2389 324.2939 14.15418466013228
1 88812U 86019    87055.95274561  .00020438  00000-0  00000-0 0 00320
2 88812 098.6990 129.3830 0160051 149.5674 211.5455 14.10001887013237
1 88814U 86019    87055.93709268  .00049871  00000-0  00000-0 0 00161
2 88814 098.9421 146.0343 0220312 233.8137 124.3125 14.85429542013806
1 88815U 86019    87055.96379351  .00043895  00000-0  00000-0 0 00150
2 88815 099.2006 147.6191 0157979 224.4236 134.4801 14.76384605013726
1 88816U 86019    87055.93498177  .00023557  00000-0  00000-0 0 00159
2 88816 098.8601 140.4803 0134664 238.4880 120.3690 14.60873007013463
1 88818U 86019    87055.96091562 -.00001072  00000-0  00000-0 0 00183
2 88818 098.5668 130.8191 0062248 138.7799 221.8659 14.21940570013185
1 88819U 86019    87053.97741826  .00157253  00000-0  00000-0 0 00124
2 88819 098.5385 128.6649 0106290 155.7053 204.9749 14.25255635012820
1 88820U 86019    87055.97110805  .00007184  00000-0  00000-0 0 00166
2 88820 098.6521 131.6270 0058129 136.6486 223.9841 14.23183435013319
1 88821U 86019    87055.98257890  .00000000  00000-0  00000-0 0 00298
2 88821 098.6655 131.5945 0095697 121.2838 239.8307 14.23138448013449
1 88823U 86019    87055.96880890 -.00021568  00000-0  00000-0 0 00440
2 88823 097.8263 117.8206 0155037 074.2384 287.6351 13.93240285012743
1 88825U 86019    87055.98508998  .00018248  00000-0  00000-0 0 00394
2 88825 098.6770 133.0001 0096517 142.2024 218.6548 14.32360407013256
1 88832U 86019    87055.99554419  .00009075  00000-0  00000-0 0 00244
2 88832 098.6855 133.3083 0065264 167.3341 193.0063 14.31341338013244
1 88835U 86019    87055.94432615  .00001346  00000-0  00000-0 0 00192
2 88835 098.5467 130.4294 0051376 128.9665 231.6660 14.21571295013178
1 88836U 86019    87055.94685682  .00005860  00000-0  00000-0 0 00181
2 88836 099.3311 139.0318 0046568 031.2002 329.2469 14.18844911012866
1 88837U 86019    87054.92861510  .00004942  00000-0  00000-0 0 00310
2 88837 097.9016 116.3456 0199511 091.2764 271.1784 13.84090857012580
1 88839U 86019    87055.94719414  .00007378  00000-0  00000-0 0 00147
2 88839 098.7618 137.4799 0093675 222.2972 137.1546 14.49281251014518
1 88841U 86019    87055.97761951 -.00001561  00000-0  00000-0 0 00099
2 88841 098.8861 136.5828 0042305 211.6567 148.2636 14.34564661013134
1 88842U 86019    87055.99434021  .00000487  00000-0  00000-0 0 00115
2 88842 098.7669 134.6236 0024840 288.8898 071.0153 14.30461691013399
1 88843U 86019    87055.93934425  .00019726  00000-0  00000-0 0 00254
2 88843 098.6889 131.6548 0077193 145.1992 215.4824 14.23208751013662
1 88844U 86019    87055.97879343  .00000631  00000-0  00000-0 0 00105
2 88844 098.6422 130.7064 0062785 097.7480 263.1379 14.16110017014175
1 88845U 86019    87055.96730543  .00000608  00000-0  00000-0 0 00129
2 88845 098.7313 130.9906 0071366 082.1249 278.8561 14.11403728013207
1 88846U 86019    87055.96073823  .00004906  00000-0  00000-0 0 00160
2 88846 099.2098 132.9498 0162087 090.7975 271.2286 13.91507793014121
1 88850U 86019    87055.94786627 -.00002246  00000-0  00000-0 0 00128
2 88850 098.6515 132.1579 0054345 130.5270 230.1215 14.24420159012630
1 88851U 86019    87055.99519553  .00014377  00000-0  00000-0 0 00347
2 88851 098.5963 131.0747 0069776 150.6184 209.9502 14.22777419013303
1 88852U 86019    87055.94964918  .00002287  00000-0  00000-0 0 00136
2 88852 098.4620 129.1963 0036973 117.2545 243.2960 14.20529397013154
1 88855U 86019    87054.97517335  .00001152  00000-0  00000-0 0 00273
2 88855 097.9742 118.5027 0194277 066.2630 295.9310 13.92317938013112
1 88857U 86019    87055.99557346  .00002690  00000-0  00000-0 0 00104
2 88857 098.5734 131.4266 0048568 148.0711 212.3987 14.25738797012996
1 88858U 86019    87055.97185193  .00013007  00000-0  00000-0 0 00137
2 88858 098.5577 131.3102 0049627 143.8374 216.6737 14.26077300012992
1 88859U 86019    87055.99223398 -.00000819  00000-0  00000-0 0 00116
2 88859 098.6008 131.3256 0051509 132.8820 227.7256 14.22701108013388
1 88860U 86019    87055.95518533  .00004315  00000-0  00000-0 0 00060
2 88860 099.0444 130.1421 0192145 097.9320 264.4211 13.86780409012634
1 88861U 86019    87055.97526721  .00054857  00000-0  00000-0 0 00270
2 88861 099.0267 141.9409 0128979 234.2760 124.6963 14.56106828013482
1 88863U 86019    87055.98096209  .00001967  00000-0  00000-0 0 00113
2 88863 098.5945 131.1010 0022775 088.2172 272.2170 14.22018417012748
1 88864U 86019    87055.93197183 -.00000244  00000-0  00000-0 0 00109
2 88864 098.7213 130.7045 0073700 082.7303 278.2782 14.10874496012560
1 88867U 86019    87054.99001258 -.00002329  00000-0  00000-0 0 00223
2 88867 098.5653 131.2282 0039839 189.0372 171.0662 14.30630781014614
```

continued...

Appendix 2 (cont'd)
SPOT 1 R/B Uncataloged Debris, February 25, 1987

```
1 88871U 86019     87055.92919871  .00065604  00000-0  00000-0 0 00326
2 88871 099.2349 133.2949 0225067 079.5962 283.1000 13.94313864012280
1 88872U 86019     87055.93093352  .00004467  00000-0  00000-0 0 00257
2 88872 098.6091 131.1616 0057888 144.7020 215.8568 14.23255425014284
1 88873U 86019     87054.97424459  .00000125  00000-0  00000-0 0 00311
2 88873 098.9305 138.0583 0092160 239.5123 119.7517 14.45640660012506
1 88874U 86019     87055.94895105  .00001055  00000-0  00000-0 0 00391
2 88874 098.7181 135.3074 0054729 246.1437 113.4578 14.40346630012618
1 88875U 86019     87055.94085874  .00002840  00000-0  00000-0 0 00120
2 88875 098.8975 137.5185 0053546 248.8571 110.7454 14.38958882012845
1 88877U 86019     87054.96323247  .00003837  00000-0  00000-0 0 00219
2 88877 098.5816 130.3071 0058467 112.8708 247.9214 14.25920074012689
1 88878U 86019     87055.98882539  .00010055  00000-0  00000-0 0 00217
2 88878 098.0049 119.6195 0178420 103.4665 258.6984 13.93236220012747
1 88879U 86019     87055.93197578  .00004003  00000-0  00000-0 0 00187
2 88879 098.6343 130.2152 0095996 128.2735 232.7671 14.15215484012607
1 88881U 86019     87055.99989605  .00001117  00000-0  00000-0 0 00156
2 88881 098.1874 123.3471 0141468 110.8093 250.8836 14.02839708012376
1 88884U 86019     87055.96480456  .00451628  00000-0  00000-0 0 00286
2 88884 099.2413 148.4433 0207221 212.0213 146.8865 14.84001253013176
1 88887U 86019     87055.96702429  .00000477  00000-0  00000-0 0 00480
2 88887 098.8993 137.9969 0070509 240.2457 119.2273 14.41263270014006
1 88888U 86019     87055.94416140  .00004942  00000-0  00000-0 0 00202
2 88888 098.6878 134.4286 0048200 232.6739 127.0616 14.36600680015971
1 88891U 86019     87055.97584995 -.00000101  00000-0  00000-0 0 00173
2 88891 098.5919 131.2110 0053365 138.9774 221.5993 14.22949834013110
1 88892U 86019     87055.96743787  .00007487  00000-0  00000-0 0 00150
2 88892 098.6472 131.6430 0038051 109.8357 250.7483 14.22803700012314
1 88893U 86019     87055.99247900 -.00133985  00000-0  00000-0 0 00277
2 88893 098.5937 130.6850 0059646 125.5935 235.1373 14.19552962012362
1 88895U 86019     87055.99118852  .00003077  00000-0  00000-0 0 00123
2 88895 099.0345 142.4403 0138527 245.0472 113.6862 14.56782464012432
1 88896U 86019     87055.97313625  .00004696  00000-0  00000-0 0 00248
2 88896 098.5742 131.1349 0047211 139.3739 221.1536 14.25553983014659
1 88897U 86019     87055.94557138  .00005424  00000-0  00000-0 0 00166
2 88897 098.5532 130.8165 0032305 131.0862 229.3672 14.24489337012720
1 88899U 86019     87055.94529076  .00003735  00000-0  00000-0 0 00116
2 88899 098.6614 132.3253 0013334 112.6274 247.6873 14.25282597012672
1 88900U 86019     87055.98561864  .00002882  00000-0  00000-0 0 00105
2 88900 098.5986 131.8628 0044573 146.5864 213.8702 14.26904766012481
1 88901U 86019     87055.98501256  .00005297  00000-0  00000-0 0 00175
2 88901 098.6377 132.0697 0029466 137.3162 223.0871 14.25118055012273
1 88903U 86019     87054.97410389  .00003947  00000-0  00000-0 0 00182
2 88903 098.6544 130.8149 0067056 119.0290 241.8181 14.22573317012966
1 88905U 86019     87055.98255872 -.00005864  00000-0  00000-0 0 00451
2 88905 098.6861 131.0958 0048895 086.5659 274.1656 14.15691651012542
1 88906U 86019     87055.96625533 -.00007024  00000-0  00000-0 0 00231
2 88906 098.7476 130.8026 0068554 097.1817 263.7700 14.10359078012775
1 88907U 86019     87055.96967597  .00001719  00000-0  00000-0 0 00145
2 88907 098.2438 124.2508 0129861 102.1466 259.4826 14.04493393012950
1 88911U 86019     87055.96771823  .00000333  00000-0  00000-0 0 00168
2 88911 098.6134 131.3517 0026427 102.8975 257.5712 14.22086230012396
1 88912U 86019     87055.99607908  .00002942  00000-0  00000-0 0 00285
2 88912 099.0810 133.1311 0143117 101.6001 260.1803 14.02238488012638
1 88913U 86019     87055.94917850  .00008808  00000-0  00000-0 0 00277
2 88913 097.9373 119.8856 0145852 064.0703 297.5950 13.99693100011912
1 88916U 86019     87055.98745924  .00001051  00000-0  00000-0 0 00144
2 88916 098.6374 132.0603 0015280 101.1291 259.2164 14.24512356011955
1 88918U 86019     87055.94727577  .00007123  00000-0  00000-0 0 00232
2 88918 098.5342 131.0427 0045368 148.1364 212.3133 14.26713714012305
1 88919U 86019     87055.99682879  .00007297  00000-0  00000-0 0 00178
2 88919 098.7006 132.9105 0037489 047.5096 312.9798 14.27487482012043
1 88921U 86019     87055.96732893  .00032598  00000-0  00000-0 0 00194
2 88921 099.4335 151.8305 0184186 250.1250 108.0628 14.86187986012757
1 88922U 86019     87055.94328270  .00079270  00000-0  00000-0 0 00157
2 88922 098.8101 144.2244 0198360 241.9608 116.2023 14.87949645012517
1 88923U 86019     87052.98486370  .00014972  00000-0  00000-0 0 00210
2 88923 099.0773 143.8755 0183933 266.8944 091.1791 14.68898509012195
1 88924U 86019     87054.99710997  .00041719  00000-0  00000-0 0 00173
2 88924 098.8330 138.8001 0107661 231.9937 127.2090 14.63999252012037
1 88927U 86019     87055.98344098  .00016548  00000-0  00000-0 0 00218
2 88927 098.8687 141.2327 0164064 247.3562 111.0805 14.62840598013245
1 88928U 86019     87054.97746401  .00005613  00000-0  00000-0 0 00144
2 88928 098.5413 129.7510 0061332 156.7189 203.7347 14.24019956012048
1 88929U 86019     87052.94800830  .00036224  00000-0  00000-0 0 00116
2 88929 098.6415 128.9479 0010792 133.2491 227.0147 14.24805612011753
```

continued...

Appendix 2 (cont'd)
SPOT 1 R/B Uncataloged Debris, February 25, 1987

```
1 88931U 86019    87055.97987419  .00167865  00000-0  00000-0 0 00367
2 88931 098.7407 143.9424 0188566 227.4984 131.0782 14.98022201013389
1 88932U 86019    87055.99294463  .00028187  00000-0  00000-0 0 00059
2 88932 098.8717 144.0019 0216756 233.7086 124.4534 14.79845380013171
1 88933U 86019    87055.98901364  .00047421  00000-0  00000-0 0 00393
2 88933 099.4024 149.9015 0210801 245.5629 112.4050 14.74573484012390
1 88935U 86019    87055.93876000  .00004772  00000-0  00000-0 0 00165
2 88935 098.2502 128.7809 0059917 305.7467 053.8720 14.33542634012417
1 88936U 86019    87055.95847044  .00002509  00000-0  00000-0 0 00126
2 88936 098.5791 129.5106 0140456 135.7320 225.5687 14.13658819012790
1 88940U 86019    87055.98934242 -.00009340  00000-0  00000-0 0 00201
2 88940 098.6700 132.9764 0014669 107.5491 252.7854 14.28726334012002
1 88941U 86019    87052.98685736  .00001195  00000-0  00000-0 0 00074
2 88941 098.7384 125.5145 0161624 105.3092 256.6531 13.95369612011740
1 88942U 86019    87054.98624189  .00000052  00000-0  00000-0 0 00253
2 88942 097.8694 117.9852 0142586 073.0406 288.6870 13.96856202011480
1 88944U 86019    87039.94733354  .00003035  00000-0  00000-0 0 00040
2 88944 099.3142 126.0851 0098197 321.2365 038.2371 14.44834952010135
1 88945U 86019    87055.97571336  .00016229  00000-0  00000-0 0 00153
2 88945 098.8800 137.1040 0027212 300.7320 059.1752 14.39881573012397
1 88946U 86019    87055.98001704  .00033944  00000-0  00000-0 0 00290
2 88946 098.8533 136.0995 0131426 230.7239 128.2809 14.43131839012278
1 88947U 86019    87055.93857821  .00005586  00000-0  00000-0 0 00162
2 88947 098.5547 130.8087 0057161 134.8316 225.8084 14.24243039011899
1 88949U 86019    87055.93810601  .00000890  00000-0  00000-0 0 00252
2 88949 098.4643 129.0987 0045770 097.7675 262.9254 14.19729545011519
1 88951U 86019    87055.94653373 -.00001194  00000-0  00000-0 0 00177
2 88951 098.6870 132.6923 0010384 070.6087 289.6770 14.24282175011693
1 88952U 86019    87055.99046253  .00005210  00000-0  00000-0 0 00314
2 88952 098.2080 127.4421 0021674 345.5359 014.5762 14.28277024011561
1 88953U 86019    87055.92999969 -.00001454  00000-0  00000-0 0 00237
2 88953 098.0788 121.0667 0164104 105.1281 256.8636 13.95975538011263
1 88954U 86019    87055.99978596  .00064743  00000-0  00000-0 0 00271
2 88954 099.0214 140.4690 0102096 259.1875 099.8385 14.46250049012088
1 88955U 86019    87054.93201372 -.00003759  00000-0  00000-0 0 00154
2 88955 098.5425 129.9463 0049150 150.2660 210.1886 14.24698376011258
1 88956U 86019    87055.96194773  .00010527  00000-0  00000-0 0 00166
2 88956 098.5962 131.4779 0085235 160.8272 199.6707 14.26068576011712
1 88957U 86019    87055.97385959  .00008222  00000-0  00000-0 0 00121
2 88957 098.5775 131.4609 0043067 128.6917 231.8685 14.26237870011991
1 88958U 86019    87055.99137070 -.00013357  00000-0  00000-0 0 00129
2 88958 098.7340 133.3524 0044130 141.1985 219.2934 14.25606659011638
1 88959U 86019    87055.95382221  .00000584  00000-0  00000-0 0 00156
2 88959 098.7121 130.6347 0075237 083.3030 277.7241 14.10638430012423
1 88960U 86019    87055.97549633  .00009128  00000-0  00000-0 0 00234
2 88960 098.9139 132.7641 0103548 112.9849 248.2830 14.12156031012432
1 88961U 86019    87055.96946768  .00002722  00000-0  00000-0 0 00204
2 88961 098.5748 131.1265 0061351 139.1277 221.5082 14.23803507014604
1 88962U 86019    87055.93503508  .00002629  00000-0  00000-0 0 00131
2 88962 098.5603 131.1722 0058889 145.6199 214.9370 14.26041623014523
1 88963U 86019    87055.98856665  .00007882  00000-0  00000-0 0 00295
2 88963 099.3321 144.8435 0165450 299.4318 059.1023 14.50681543011479
1 88965U 86019    87055.97269407 -.00001478  00000-0  00000-0 0 00261
2 88965 098.4775 130.1726 0047192 147.5640 212.9014 14.24627593014465
1 88966U 86019    87055.98446209  .00000741  00000-0  00000-0 0 00216
2 88966 098.4525 129.4745 0032068 138.7923 221.6243 14.22818089014239
1 88967U 86019    87055.96407500  .00004024  00000-0  00000-0 0 00102
2 88967 098.6666 133.5613 0096387 168.0094 192.3979 14.32142645013820
1 88968U 86019    87055.97250875  .00007958  00000-0  00000-0 0 00250
2 88968 098.6284 132.0201 0046074 123.2599 237.3564 14.25692597012181
1 88970U 86019    87055.94208223  .00000106  00000-0  00000-0 0 00191
2 88970 098.9607 130.9168 0151361 070.7042 291.0950 13.95173752011874
1 88973U 86019    87053.95590077  .00006654  00000-0  00000-0 0 00110
2 88973 098.6479 132.9949 0130317 311.3637 047.7008 14.42211333011390
1 88974U 86019    87055.96506543 -.00002606  00000-0  00000-0 0 00135
2 88974 098.6291 131.3911 0023981 120.6017 239.8085 14.21723193013172
1 88975U 86019    87055.97503884  .00000562  00000-0  00000-0 0 00107
2 88975 098.5895 131.6892 0031023 147.5282 212.8373 14.25886251012990
1 88977U 86019    87052.96442461  .00002002  00000-0  00000-0 0 00031
2 88977 099.0702 138.2060 0109769 275.7236 083.2023 14.48863048010737
1 88978U 86019    87055.98785296  .00003173  00000-0  00000-0 0 00145
2 88978 098.6951 132.2024 0027349 074.8161 285.6593 14.22187439010950
1 88979U 86019    87043.93204673 -.00080471  00000-0  00000-0 0 00049
2 88979 098.7588 119.2973 0063122 147.5044 213.0588 14.12341515009473
1 88980U 86019    87055.99588829  .00000255  00000-0  00000-0 0 00106
2 88980 098.6644 130.6378 0061590 087.4272 273.4498 14.13933270011182
```

continued...

Appendix 2 (cont'd)
SPOT 1 R/B Uncataloged Debris, February 25, 1987

```
1 88982U 86019     87055.99813805  .00004913  00000-0  00000-0 0 00209
2 88982 098.6830 133.0987 0008624 208.9216 151.2050 14.29007714011090
1 88983U 86019     87055.96403399  .00005294  00000-0  00000-0 0 00337
2 88983 098.4055 129.4392 0022713 342.5786 017.5173 14.26668138010975
1 88984U 86019     87055.94174772 -.00002634  00000-0  00000-0 0 00242
2 88984 099.0570 136.7230 0100958 346.0528 013.8418 14.23519681012325
1 88986U 86019     87055.97167887  .00006494  00000-0  00000-0 0 00423
2 88986 098.7327 136.0610 0079430 245.2474 114.1008 14.43111486012343
1 88987U 86019     87055.96358607  .00018107  00000-0  00000-0 0 00250
2 88987 098.4777 130.6137 0035569 133.5576 226.9129 14.29400162010857
1 88988U 86019     87055.97103672  .00009893  00000-0  00000-0 0 00254
2 88988 098.5441 131.0331 0048638 155.2317 205.1773 14.26320776010823
1 88993U 86019     87053.95187026  .00005301  00000-0  00000-0 0 00127
2 88993 098.7180 129.3135 0084446 099.2863 261.8418 14.14769448010402
1 88994U 86019     87055.94203308  .00005173  00000-0  00000-0 0 00117
2 88994 098.5508 131.4005 0054874 164.9439 195.3950 14.27691589014473
1 88995U 86019     87055.94858998  .00001158  00000-0  00000-0 0 00196
2 88995 098.5786 131.7613 0015160 155.9193 204.3259 14.27308195013472
1 88996U 86019     87055.95781705 -.00006594  00000-0  00000-0 0 00177
2 88996 098.6627 131.7465 0027036 118.2366 242.2067 14.21979424014430
1 88997U 86019     87055.95404617  .00003724  00000-0  00000-0 0 00094
2 88997 098.8134 120.5576 0419245 097.1594 267.7863 13.40332055010592
1 88998U 86019     87055.96140233  .00000889  00000-0  00000-0 0 00221
2 88998 098.5183 129.3373 0091618 137.5420 223.3442 14.17246007014594
1 88999U 86019     87055.95102933  .00000661  00000-0  00000-0 0 00101
2 88999 098.6846 130.7653 0063208 082.0846 278.8041 14.13622615014385
```

Appendix 3
NAVSPASUR Notification of Fragmentation

ROV MSG	TIME	PADAY	IMMEDIATE
76945	2336	318/86	

WHEN AN ADDRESSEE CLAIMS NONRECEIPT OF A MESSAGE YOU SENT, YOU SHOULD PROVIDE THE DATE, TIME, GROUP TO THE ADDRESSEE SO THEY CAN REQUEST A COPY OF THEIR SERVICING TOO. IF THIS FAILS, PLS. USE THE AF FORM 1026 TO REQUEST A RETRANSMISSION.

O 1423137 NOV 86
FM NAVSPASUR DATA, DAHLGREN VA
TO RUNTNOAL/JSS PACECOM PETERSON AFB CO/TBE/J3SS//
BT
UNCLAS//NO 3840//
FOR HQ SPACECMD CODE J3SC, J5YS,TBE

SUBJ: SATELLITE SUPPORT
SUBJ: BREAKUP ANALYSIS
1. ANALYSIS OF THE BREAKUP OF SCC NO. 16615 YIELDS THE FOLLOWING TIME AND LOCATION OF BREAKUP BASED ON ELEMENT SETS GENERATED ON 17 ASSOCIATED PIECES:

DATE/TIME	LAT(N)	LONG(E)	ALT (NM)	STDEV (NM)
861113 193923.2	7.0	42.0	435.2	1.0

BT
#6618

Appendix 4
Acquisition and Correction of SPOT 1 Orbit

IAF-86-228

ACQUISITION AND CORRECTION OF SPOT 1 ORBIT

Brigitte SERREAULT

CNES 18 Avenue Edouard Belin 31055 TOULOUSE CEDEX

ABSTRACT

SPOT has a quasi circular, sun synchronous orbit with a descending node at 10 : 30 AM.

In order to assume a systematic image acquisition all over the earth, this orbit has a 26 days cycle with a 5 days sub-cycle. The 369 theoretical ground tracks of a cycle define a grid (with 369 equidistant descending nodes on the equator). The satellite must "reach" the nearest orbit of the grid and remain in a 3 km cross track at the equator and 5 km above 60° window.

This constraint involves 3 types of orbital maneuvers : rendez-vous, inclination and small phasing corrections.

This paper describes the initial orbit acquisition (rendez-vous without inclination corrections, owing to the very good initial orbit delivered by the launcher), the strategy and realization of the phasing corrections since injection (evaluation frequency and obtained results) and the estimated next maneuvers.

KEYWORDS

Spot - sun-synchronous satellite - orbitography - orbit acquisition - orbital maneuvers - orbital perturbations - extrapolation - phasing -

INTRODUCTION : SPOT ORBITAL CHARACTERISTICS

SPOT1, which was launched on February 22nd 1986, is the first one of a series of sun synchronous satellites deviced to remote sensing. The program is expected to last 10 to 12 years, and is based on one to two satellites in flight at the sametime, with a supporting ground system (including a Control Center, a Mission Center and the Orbitography System).

The orbital characteristics (orbital parameters, orbit acquisition and corrections) have been determined by the objectives of the mission and the payload limitations (HRV resolution and swath width) which are :

. good resolution without scale fluctuations : which leads to choose a low quasi circular orbit (but with a non zero eccentricity)

. constant lighting with good solar elevation (avoiding specularity) : which leads to a sun synchronous orbite with a local time close to 11 hours or 13 hours, at the descending node (in order to get the greatest luminosity over the northern hemisphere)

. global coverage of the earth (which leads to a polar phased orbit : inclination near 90 degrees)
Considering the HRV swath width (60 km/instrument, 117 km for the both) which leads to a cycle greater than 25 days (the distance between two adjacent nodes on the equator must be smaller than the overall swath width)

. stereo coverage (using the mirrors) with a short period of time : (definition of "sub-cycles")

The mean orbital parameters, resulting from a compromise between all these specifications, are summarized in the following table :

semi major axis	: a = 7 200.55 km
eccentricity	: e = .00106
inclination	: i = 98.723°
perigee argrement	: w = 90°
right ascension of the node	: Ω = 77.32° + .985612*DJ (DJ = julian day since january 1st 1950)

continued...

Appendix 4 (cont'd)
Acquisition and Correction of SPOT 1 Orbit

These parameters are identical for all the satellites of the series. Il there are two satellites at the same time in flight the second will be at 50 minutes (half a period) behind, on the same inertial orbit (with a difference of mean anomalies of 180 degrees) (to avoid problems for the ground stations)

In term of mission, they are equivalent to the following data :

revolutions/day	: 14 + 5/26
orbital cycle	: 26 days
orbital "subcycle" (time of access to the adjacent descending node on the equator towards west)	: 5 days
number of tracks/ orbital cycle	: 369
intertrack distance:	108.4 km (on the equator)
mean local time	: 10:30 am at descending orbit 10:30 pm at ascending orbit

The orbital symmetry leads to define a mean theoretical grid of 369 orbits (symmetrical around the polar axis), with 369 equidistant descending nodes on the equator. (the 370th node would be identical to the 1st one). The 1st orbit which is reproduced 368 times (after a rotation of the node) corresponds to a mean orbit defined with zonal terms and middle of life parameters. This grid has been fixed before launch in order to facilitate the preparation of the Spot Image catalogue, the mission center preparations and the stations designation.

Its 1st node is at the longitude of 330.24 deg (East).

The grid and the swath width fix the orbit maintain and acquisition :

with such an intertrack distance (108.4 km) compared to 117 km for the HRV, it "remains", on the equator, 6 km for the orbital deviations. At 60° of latitude, the tracks are closer ; it "remains" 10 kms.

so, the orbit must reach one of the track of the grid, and remain in a window of more or less 3 kms around this theoretical track at the equator and 4.5 kms at 60° of latitude.

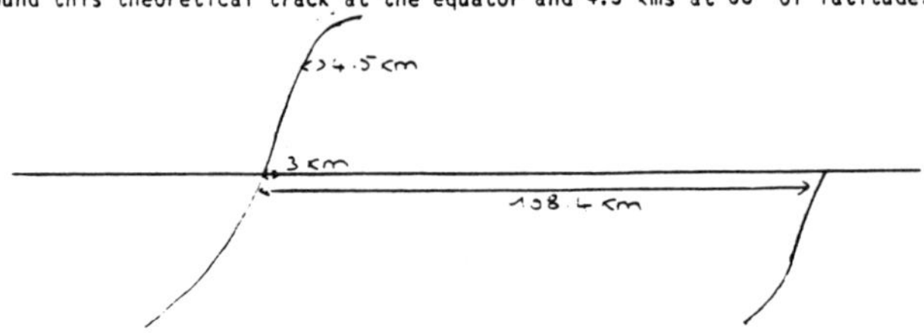

All the orbit corrections, acquisition and quality depend on this highly constraining specification.

We shall see now what makes this orbit "leave" ist window and what is the strategy to reach and stay around the grid which has been applied for SPOT1.

NATURAL ORBITAL EVOLUTION

There are 3 causes : atmospheric drag, luni solar potential, and of course geopotential.

Geopential

If is perfectly predictible and has been taken into account when defining the grid and the analytical extrapolation and numerical restitutions models (the secular terms define the sun synchronous phased orbit).
The only pertubation may come from the long period perturbations on the eccentricity and the argument of perigee (which must be as constant as possible for the image processing).

If ex is e*cosw, ey is e*sinw, and e the vector (ex, ey), due to the zonal terms (J2 and the odd zonal terms : J_{2l+1}), this vector will oscillate around the direction ey (w = 90°) and if the eccentricity gets too small, turn and create altitude differences above a fixed geographical point.

continued...

Appendix 4 (cont'd)
Acquisition and Correction of SPOT 1 Orbit

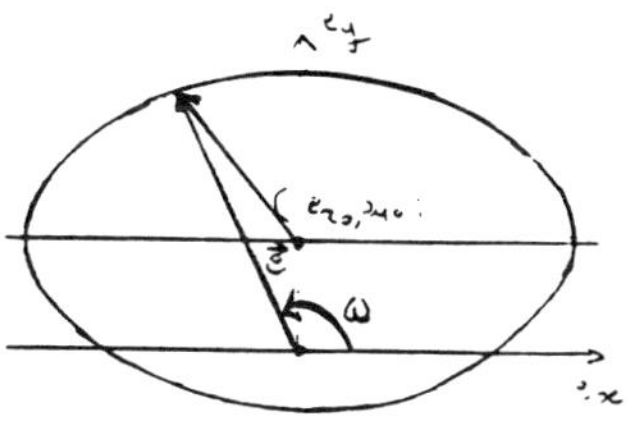

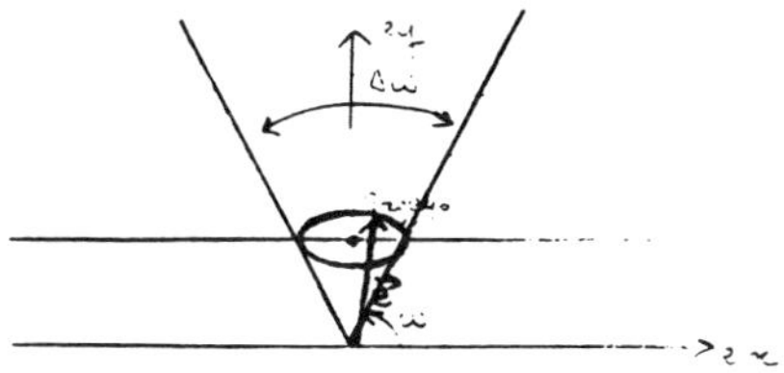

The ellipse center is named the "frozen point". The eccentricity will be kept as close as possible of this point ($e \simeq 1.0610^{-3}$ for SPOT, $w = 90°$). It will be systemmatical corrected if the distance of the vector to the frozen point increases, by maneuvers along the roll.

Luni-solar potential

Due to the sunsynchronism the relative sun-earth-satellite geometry is nearly constant. If αs^* is the mean sun right ascension,

$(\alpha_s = ws + M_s$
$\simeq ws + \Omega_s + M_s)$
and Ω^*, the mean value of SPOT Ω at the day DJ :
$\alpha_s^* - \Omega^* \simeq 202.5°$
which is constant

The solar attraction straightens up the inclination i, with a secular rate :

$$\frac{di}{dt} \simeq -\frac{3n_s^2}{n} \sin i \, . \cos\left(\frac{i_s}{2}\right) . \sin 2\left(\alpha s^* - \Omega^*\right)$$

i_s is the inclination of the eccliptic on the equator

n_s and n are the solar and satellite mean motions

$$\frac{di}{dt} \simeq -0.915 \; 10^{-4} \text{ degree/day}$$

This is a mean effect which is complicated by long (450 days) and short (13.5 days) effects on the inclination and the right ascension of the node due to the lunar potential, and a parabolic decrease of Ω ($d\Omega/dt = f(i)$)

Within 6 months, the inclination decreases of $-1.647 \; 10^{-2}$ degrees, and the local time varies of 42 seconds.
The effect of the inclination increases with the latitude (it is zero at the node)

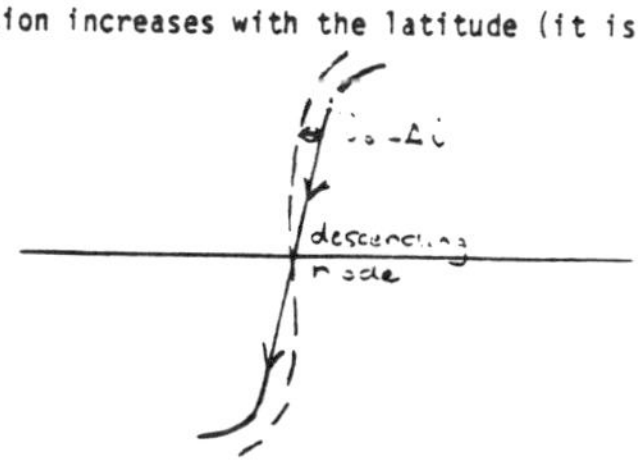

6 months decrease means 1 km crosstrack at 60° latitude. This must be corrected by maneuvres along pitch.

Atmospheric drag

It is the most annoying force, it is non repetitive and uneasily predictible for it is completely dependent on the atmospheric density and the solar activity. Except in a magnetic stormy period, the effects are nearly constant on the orbit and during a day.

The mean result is a secular decrease of the semi major axis and, as a consequence, of the nodal period.

Roughly : if P is the nodal period

$$P \simeq \frac{2\pi}{n} \text{ with } n = \mu/a^3$$

continued...

Appendix 4 (cont'd)
Acquisition and Correction of SPOT 1 Orbit

(n = mean motion
μ = gravitation coefficient
a = mean semi major axis)

so $= \frac{\Delta P}{P} \sim \frac{3\pi}{2} \frac{\Delta a}{a}$

$$\Delta P \simeq 3\pi \Delta a \sqrt{\frac{a}{\mu}}$$

If the nodal period decreases of ΔP, the longitude of the node changes :

$\Delta L = \omega t . \Delta P$ (ωt = earth rotation)

After n orbits = $L_n - L_{n\ nominal} = \Delta L1 + \Delta L2 + \ldots \Delta Ln$

$$\Delta L(n) = \omega t . \frac{n(n-1)}{2} . \Delta P$$

The effect, for a secular decrease of a, is parabolic. For a given inclination, there is a specific value of semi major axis for which the nodes don't shift (the mean period is perfectly nominal).

This value will be named "nominal", ao.

If the a value is greater : the descending nodes shift from East to West
until a reaches ao
If the a value is smaller : the descending nodes shift from West to East

MODIFICATION OF THE ORBIT

Injection

Due to the launcher there is a dispersion on each parameter, and the initial track has a low probability to be in the window.

Safe mode

In case of emergency, if all the preprogrammed reconfigurations fail, the spacecraft enters a safe mode after changing the earth pointing for a spinned sun pointing mode. In this mode, the on board-computer is off line. The piloting is assumed by the thrusters instead of the reaction wheels, which produce non neglectible thrusts. The satellite looses the phasing.

Orbital corrections

While thrusting, piloting is assumed by thrusters if the maneuver is important (MCO called mode), by reaction wheels if not (MCC mode). In MCO mode, for big maneuvers the perturbations get significant (inclination corrections) and disturb the phasing.

STRATEGY

Orbit maintenance

It results from a compromise between all the previously described perturbations :

- the inclination decrease (maximum effect at high latitude) and
- the nodes shift (maximum effect on the equator)

considering the atmospheric drag prediction imprecision (taking in account a tolerance margin)

. With an annual inclination correction, it remains 2 km for the nodes shift at the equator, and 1 km at 60° ot latitude which is coherent

so : the nominal mean inclination is i_{nm} = 98.723°

the initial is i_0 = i nom + (Δi/day) * 183 days = 98.739°

The annual maneuver will be decided at i_f = 98.707°
to reach i_0

. The inclination corrections are difficult : there are no 15.6N thrusters along pitch direction : the spacecraft must turn around yaw before thrusting

continued...

Appendix 4 (cont'd)
Acquisition and Correction of SPOT 1 Orbit

Semi major axis, eccentricity and argument of perigee are corrected at the same time, by small maneuvers of 2 thrusts, when the satellite is in the night or over the sea (to avoid image collecting perturbations) the frecency depends on the solar activity;
Most of these corrections should be made in the MCC mode (short correction) (Δ a varies from 20 to 150 meters)

Orbit acquisition and reacquisition

The phasing is reached by "rendez-vous" maneuvers when the satellite is out of the window .
If the node shift is important there will be one correction ; when the node reach by themselves, the east margin of the grid "return rendez-vous".
If the natural shift seems to long, there will be a preliminary correction ; "going rendez-vous", taking into account the drift duration and the amplitudes ot the thrusts.
In acquisition, the choosen position in the grid is the nearest.
In re-acquisition, it is fixed (ex : SPOT2)

SPOT 1 : CHRONOLOGICAL ACCOUNT OF EVENTS - RESULTS

Initial acquisition

The orbit delivered by the launcher was nearly the specified initial one :

Δ a = 130 m
Δ e = 2.10^{-4}
Δ i = - 3.810^{-3} deg
Δ w = 0.737°
$\Delta\Omega$ = 2.3 10^{-4} deg

The inclination appeared so close the i_o value, that is was unecessary to correct it.

The 1st tracks placed 11 km west the grid (8 km from the window), with a natural slow drift towards west (semi major axis above the nominal : Δ a = 1.18 km). The next return to the margin would have taken place on J + 11 days.

If was decided to reach the nearest east margin at J + 7 (2 : PM) to phase earlier the tracks, to test and calibrate the thrusters, and to correct e and w with a "going" maneuver.

expected corrections : Δ a = 1.16 km
Δ e = - 2.10^{-4}
Δ w = 4.39°
Consommation = .88 kilos

(latitudes and Δ V are imposed :

1st thrust : 61.77° (ascending orbit) DT = 13 seconds on roll minus axis
2nd thrust : 50 minutes later : (-61.78° descending) DT = 56 seconds on roll plus axis)

These "long" corrections are realized by boards of 15.6 newtons each thrusters (two pairs on each side along roll) (SPOT mass is around 1 835 kilos). Elementary impulses of the MCO mode reach 1 second (about 33 meters on a for 1.0 second, BOF)

The values were obtained with the following precision :

Δ a $\simeq$ -21 m
Δ e $\simeq$ 7.10^{-6}
Δ w $\simeq$ .049°

The differences came from the thrusters stabilization during the maneuvers, and appear like a 98.5 % global efficiency (obtained ΔV = 0.5311 m/sec)

The orbit was adjusted by 4 stations (with range and range-rate measurments)

"Return" maneuver took place at 2.4 km east from the grid, and was computed for reaching the nominal semi major axis plus 10 m (to compensate the predicted atmospheric drag).

Expected Corrections : Δ a = -1.88 km
Δ e $\doteq$ 3.910^{-6}
Δ w = .05°

The position of the 1st thrust is free

1st thrust (over KURU) (10° lat descending) = 36 sec on roll + axis
2nd thrust (10° lat ascending) = 35 sec

Consommation = 0.8 kilos

continued...

Appendix 4 (cont'd)
Acquisition and Correction of SPOT 1 Orbit

The differences with the expected values were :

Δa = 94 m
Δe = $5\text{-}10^{-7}$
$\Delta\omega$ = 0.05°

and were issued of the stabilization (rounded to a second and dissymetric pulses)

After restitution of attitude control, the efficiency proved to reach again 98,5 %.

During these two maneuvers, attitude errors remained under 1° around each axis (.6° around roll) and normal pointing was reestablished 60 seconds after the last thrust pulse. These maneuver are programmed with an interactive task (considering stations visibilities, optimal drift and consommation). Return maneuver was automatically computed.

Final phasing acquisition (MCC maneuver)

With a too high semi major axis (due to the dispersions), the nodes kept on drifting towards west slowly. When reaching, the western margin of the grid, a maneuver was programmed, to decrease the semi major axis and correct a and w, with the automatic phasing maintain programming. (expected a is near the "nominal" value)

After correction, nodes were expected to go east and read the margin within 4 weeks (Δa = - 105 m) (see diagram below)

During this maneuver global attitude error never reached .5°. Semi major axis was reached at 2 meters (error due to the pulse rounding : .0625 second). The .985 thruster calibration was used for computing (see method below).

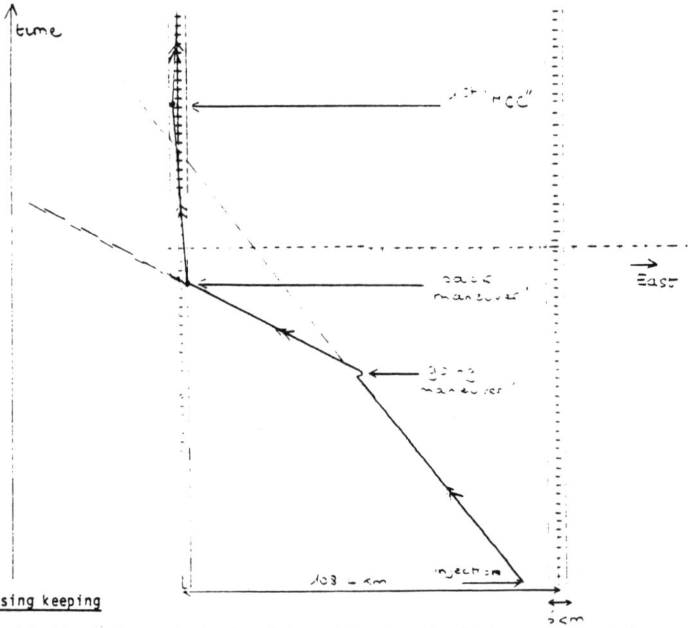

Phasing keeping

The objective is to remain in the window, while the nodes follow a mean parabola between the margins.

The problems are :

- to measure a mean (without periodic perturbations) semi major axis
- to have a good orbital adjustment quality to measure a small daily decrease (Δa/day $\simeq$ .30 m)
- to get a good restitution of the atmospheric drag effects and to predict it up to 2 months
- to predict correctly the extrapolated nodes up to 2 months : error along the track inferior to .5 km (for the post maneuver parameters until the next maneuver).

continued...

Appendix 4 (cont'd)
Acquisition and Correction of SPOT 1 Orbit

The orbit computation is made with an analytical specific extrapolation model. The mean elements are adjusted twice a day after the numerical orbit adjustment ; they are the six classical elements, and two coefficients : a drag coefficient and a fixed scale factor.
Drag is extrapolated after averaging twice the daily semi major axis decrease.
Considering the predicted drag, the maneuver is computed in this way ; after extrapolations and research, we get a1 = the mean semi major axis to reach the western margin until the next predicted orbital correction and a2, the semi major axis to reach the eastern margin, the a value, expected after corrections is : $a = (a_1 + a_2)/2$

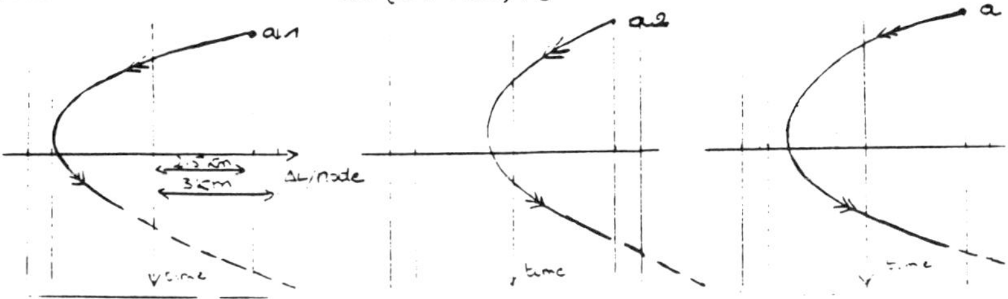

The period between 2 corrections depends on the solar activity. It varies from 3 to 10 weeks (value computed from the estimated drag coefficient).

If the eccentricity vector reaches the authorized limit ($\Delta w = 13^\circ$), the maneuver computation takes the correction into account (to come back to the ellipse center). Such a maneuver would be realized in MCO mode (long thrusts) with rounded to a second pulses. To avoid these corrections, all the phasing maneuvers have corrected "slightly" the eccentricity vector (within the limits of the MCC mode) after an interactive preparation (see schema below)

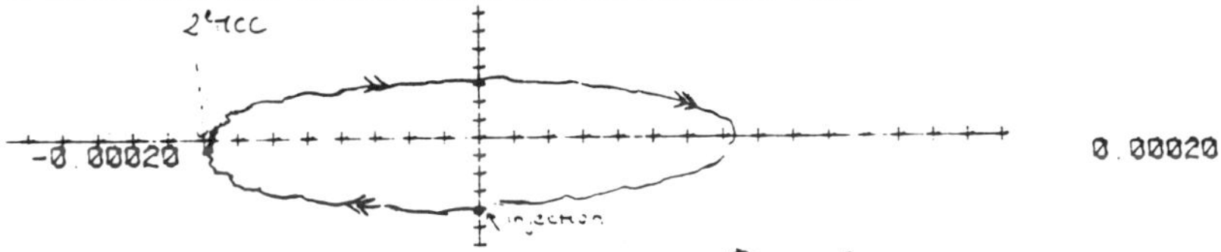

Results : for a mean aimed value of 20 m for Δ a (with $[\Delta \vec{e}] \simeq 10^{-5}$) the time between corrections has reached 4 months, without going out of the margins. (schema)

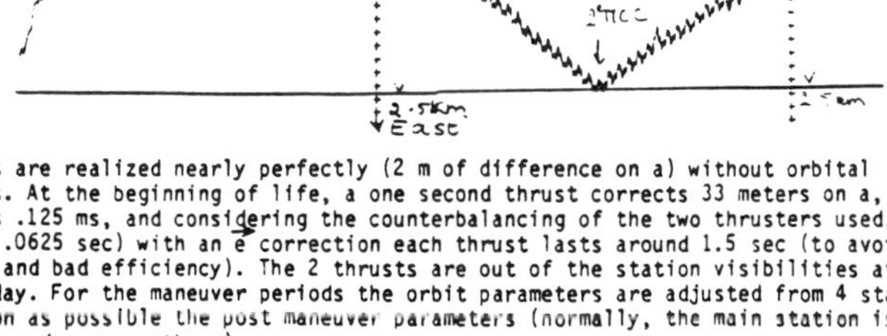

The maneuvers are realized nearly perfectly (2 m of difference on a) without orbital perturbations. At the beginning of life, a one second thrust corrects 33 meters on a, the elementary pulse is .125 ms, and considering the counterbalancing of the two thrusters used, the rounding reaches .0625 sec) with an $\vec{e}$ correction each thrust lasts around 1.5 sec (to avoid too short pulses and bad efficiency). The 2 thrusts are out of the station visibilities at the same hour in the day. For the maneuver periods the orbit parameters are adjusted from 4 stations : to get as soon as possible the post maneuver parameters (normally, the main station is TOULOUSE, with one pass a day over others).

For a weak solar activity period (as in 1986), the major difficulty is the long termprediction and the quality of the adjustment (reaching the limits of the geopotential model).
Mean parameters follow the expected behavior. Nearly descending straight lines for a, i, and , ellipse for the e vector, without big perturbations due to the variations of the solar activity. With greater perturbation it is sufficient to reduce the period between maneuvers and the window around the grid.

continued...

Appendix 5
Letter from Donald Kessler, NASA, to Col. Stephen Marlow, USSPACECOM, December 1, 1986

National Aeronautics and
Space Administration

Lyndon B. Johnson Space Center
Houston Texas
77058

NASA

SN3-86-249 December 1, 1986

Col. Stephen Marlow/J3SOT
Hq. US Space Command
Colorado Springs, Co. 80914

Dear Col. Marlow:

As you are aware, the 3rd Stage of the Ariane launch of 22 February, 1986 broke up into an estimated 200 plus trackable fragments. NASA has already informally expressed concern about this breakup to both the European Space Agency (ESA) and to the builder of the Ariane, and offered assistance in determining its cause. We plan a more formal working relationship once the appropriate personal are identified by the ESA. Once a cause has been established, it is hoped that corrective actions can be taken.

One of the issues to be resolved is whether the breakup was the result of an explosion (hence correctable by ESA) or a collision (requiring corrections by all of the space community). From other measurements, we know that the smaller fragments from a collision are more numerous and have higher ejection velocities than the fragments from an explosion. In addition, the fragments are darker. Consequently, we are asking for support in the following areas to help resolve the cause of this breakup.

1. Record about 90 minutes of Eglin radar data with the fence down as the orbital plane passes over Eglin, as was done for the Delta 180 project.

2. Record about 90 minutes of PARCS radar data using the "Small Satellite Software" as the orbital plane passes over PARCS.

3. Obtain IR measurements (approximately 2 hrs/day for 10 days) using the AMOS and MOTIF telescopes on selected fragments. These observations would require Kaena Point radar support to provide pointing information for the telescope and to to obtain RCS data. We plan to obtain visible data on the same fragments using NASA telescopes. We will coordinate these observations with AMOS, MOTIF, and Kaena Point.

4. If there are any instruments in space which we are not aware, such as optical (visible or IR) or radar, which could detect even smaller objects, we would be interested in obtaining piece counts of fragments as this instrument passed near the orbital plane of the Ariane fragments. To be effective, any instrument would have to be capable of detecting objects with high angular

continued...

Appendix 5 (cont'd)
Letter from Donald Kessler, NASA, to Col. Stephen Marlow, USSPACECOM, December 1, 1986

velocities passing within a few 10's of kilometers. Any optical instrument should be very sensitive to low light levels.

In addition, some data analyzed by Teledyne Brown Eng. suggest that this rocket stage may have a history of earlier breakups. Earlier launches used the 3rd stage as a transfer stage to geosynchronous orbit and were left in approximately 300 km by 36000 km, 7 deg. inclination, orbits...orbits which are difficult to detect and catalogue. In order to test this possibility, we need the following additional data:

5. Record about 90 minutes of Eglin radar data with the fence down as the perigee of one these transfer stages passes to the south of Eglin.

Exact times of the tests can be coordinated with Ron Kling or Dave Nauer of Teledyne Brown Eng. If you have any questions, contact me at 713 483 5347

Sincerely,

Donald J. Kessler

Donald J. Kessler

Appendix 6
Memorandum from G. Taft DeVere, TBE, to William P. Stillman, NORAD, January 31, 1984

TELEDYNE
BROWN ENGINEERING

MEMORANDUM

TO: William P. Stillman, XPYW **DATE:** January 31, 1984

FROM: G. Taft DeVere

SUBJECT: Breakup of 1979-104, Ariane V1

As required by contract F05604-82-C-0061 SOW part 3, chapter 2 para 2c(2) and the tasking letter dated 15 Feb 83, attached is our analysis of a recently discovered satellite breakup. Please direct any questions to the undersigned at 574-7270.

G. Taft DeVere
G. Taft DeVere
Systems Analyst

GTD/cb

cc: Preston Landry, XPYS (2)
Pat Jardin, 1014 CONS/LGCSA
(w/o attch)

continued...

Appendix 6 (cont'd)
Memorandum from G. Taft DeVere, TBE, to William P. Stillman, NORAD, January 31, 1984

In our continuing analysis of satellite breakups and 8XXXX satellites, we have uncovered a previously unknown breakup. Although data on this event is somewhat sketchy, below is listed our interpretation of this event. If further information becomes available, we will keep you apprised.

On December 24, 1979 the French launched their first Ariane booster from Kourou, French Guiana. This did not contain an important payload, but was designed as a booster test. The orbit was approximately inclined 17.65° with an apogee of 36,000 km and a perigee near 200 km. Shortly after launch, we have reason to believe that the rocket body, object 11659, exploded due to a closed vent valve. Over the next few months, several different 8XXXX satellites entered and left the SCC data base. Sometime during the first week of May, 1980 object 11792, alleged to be the upper stage rocket body of Intelsat 2F-1 (launched in 1966), was cataloged. It appears that this object is associated with the Ariane breakup. Various orbital information for the last 4 years was examined, and an analysis of right ascension showed that most pieces converged on the launch date, with some converging around day 130 of 1980. This may support the theory of a second breakup. A total of 37 8XXXX's were extracted from various CLASSY and SCC/NSSC listings while some of these may be the same objects that were deleted and later rediscovered, we feel that at least 25 of these are unique. A Gabbard gram is attached. Due to poor surveillance at the low inclination of 17°, many of these objects have been lost. As of 16 Jan 1984, there were only Five 8XXXX's in the catalog which are associated with Ariane. They are:

80058
80059
83715
85655
85710

We suggest that particular care be placed in processing these satellites and the Eglin UCTs which are at low inclinations. These objects should be tasked to other sensors and incorporated into the catalog as soon as possible. Precession of perigee of these orbits will take them in and out of the coverage of the FPS-85 for days or weeks at a time, so tasking these to other sensors is important if these type of objects are to be cataloged. Also, an attempt should be made to find the correct rocket body for Intelsat 2F-1, which should have an orbit similar to its payload, object 2514 (The perigee of object 2514 is approximately 3,200 km, more than 10 times that of any of the Ariane pieces). When the correct Intelsat rocket body is found, its elements should be substituted for those of 11792, whose elements should be recataloged as a piece from 1979-104.

continued...

Appendix 6 (cont'd)
Memorandum from G. Taft DeVere, TBE, to William P. Stillman, NORAD, January 31, 1984

TELEDYNE
BROWN ENGINEERING

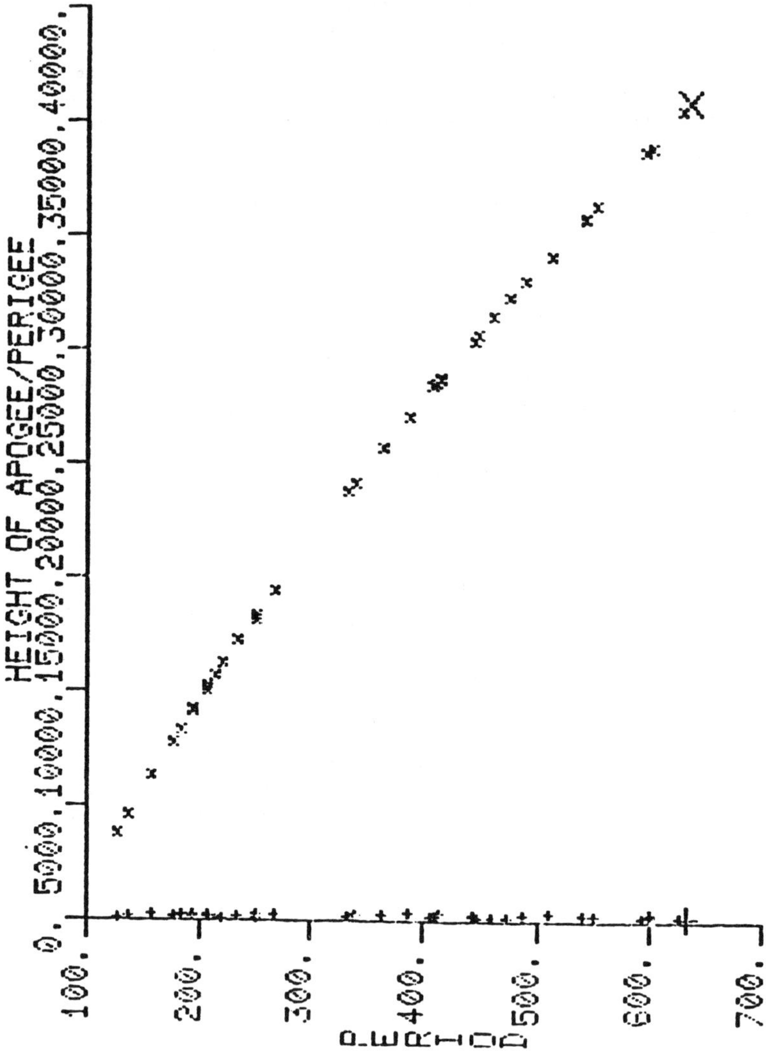

Distribution of Pieces 1979-104 Ariane V1

Appendix 7
Letter from I. W. Pryke, ESA, to Nicholas Johnson, TBE, May 8, 1984

european space agency
agence spatiale européenne
europäische weltraumorganisation

MR NICHOLAS JOHNSON
TELEDYNE BROWN ENGINEERING
1250 ACADEMY PARK LOOP, SUITE 240
COLORADO SPRINGS, CO 80910

REF.: 214/2712
MAY 8TH 1984

DEAR MR JOHNSON,

WHEN WE MET EARLY IN APRIL YOU GAVE ME A COPY OF THE ATTACHED PLOT, RELATING TO DEBRIS FROM THE ARIANE 1 LAUNCH.

I HAVE MADE ENQUIRIES IN EUROPE AND AM INFORMED THAT L01 DID CREATE A LARGE NUMBER OF DEBRIS. ACCORDING TO THE MILLSTONE RADAR TEAM THE THIRD STAGE BLEW UP APPROXIMATELY 90 DAYS AFTER LAUNCH. THIS WAS THOUGHT TO HAVE OCCURED AS THE FUEL AND OXIDENT VALVES ON THE STAGE WERE DELIBERATELY LEFT CLOSED FOLLOWING LAUNCH (ON SUBSEQUENT FLIGHTS THE VALVES WERE LEFT OPEN AND THIS COULD ACCOUNT FOR THE DIFFERENCE IN AMOUNT OF DEBRIS). HOWEVER, MILLSTONE STATED AT THE TIME THAT THE DEBRIS WERE IN AN ORBIT WITH A PERIGEE OF APPROXIMATELY 90 MILES THAT WAS RAPIDLY CIRCULARISING (PERIOD 110-120 MINUTES) WITH RE-ENTRY EXPECTED TO OCCAR RAPIDLY. THIS DOES NOT SEEM TO BE CONSISTENT WITH YOUR PLOT.

IF YOU WISH US TO REVIEW THIS MATTER FURTHER OUR ARIANE OFFICE IN PARIS REQUESTS THE FOLLOWING INFORMATION:

- THE ORBITAL INCLINATION OF THE DEBRIS
- THE EPOCH (TIME OF ORBIT ESTIMATE)
- NATURE OF DEBRIS (RADAR CROSS SECTION).

I HOPE THE ABOVE PROVES OF USE TO YOU.

YOURS SINCERELY,

I. W. PRYKE
HEAD, WASHINGTON OFFICE

Appendix 8
Telegram from Max White, Royal Greenwich Observatory, to Nicholas Johnson, TBE, November 24, 1986

TDYBRWN HTV

87451 RGOBSY G
TDYBRWN HTV

5350 86-11-24 17:02

TO: NL JOHNSON TELEDYNE BROWN ENGINEERING COLORADO SPRINGS
: SVEN GRAHN SWEDISH SPACE CORP STOCKHOLM
FM: MAX WHITE SATELLITE TEAM RGO UK

SOME INITIAL DATA FOR YOU ON ARIANE FLASH PERIOD. WE HAVE YET TO ANALYSE PLATES TAKEN BY CAMERA. BUT DOUBT THAT WE COULD USE THESE ANYWAY. AS THE FIELD OF VIEW WOULD NOT SHOW ALL THE FLASH PERIOD IN AN EXPOSURE, WITH BEING STATIONARY AND NON TRACKING.

THESE ARE TAKEN FROM VISUAL OBSERVATIONS MADE IN EDINBURGH AND WEST GERMANY (E/G). STILL AWAITING LATER DATA FROM GERMANY FOR THE LAST 2 MONTHS. MOST OF OUR OBSERVATIONS REPORTED ON 86 19C SHOW A REGULAR OR IRREGULAR FLASH PERIOD, BUT UNFORTUNATELY THE DURATION OF THE FLASH CYCLE HAS NOT BEEN WELL COVERED.

MARCH 18/25/30 : FROM E - REGULAR VARIATIONS EVERY 50 SEC
APRIL 23 : FROM G - FLASHES TIMED AT 66 SECONDS
APRIL 26 : FROM E - TIMED AT 50 SEC REGULAR VARIATIONS
MAY 2 : FROM G - TIMED AT 65 SEC
MAY 3 : FROM G - TIMED AT 53 SEC
SEPT 7 : FROM E - TIMED AT 30 SEC REGULAR VARIATIONS

THE GERMANY OBSERVATIONS WERE DESCRIBED AS TUMBLING. NO SHARP FLASHES WERE OBSERVED FROM EITHER THESE STATIONS OR OTHERS. BUT ADDITIONAL VISUAL OBSERVATIONS REPORTED DESCRIBED THE APPEARANCE AS IRREGULAR FLASHES. WHEN HAVING SEEN THE OBJECT MYSELF OBSERVING HERE. I WOULD SAY THAT THE FLASHES WERE OF REGULAR DURATION. AS IF THE OBJECT WERE DOING A ROLL MOTION AND NOT AN END OVER END MOTION - HOWEVER RADAR IS MORE SUITED FOR DETERMINING THIS, AND I AM THINKING BACK A FEW MONTHS.
I WOULD SAY THAT THE DISCREPANCIES IN SOME OF THE TIMINGS AROUND THE SAME PERIOD MAY BE DUE TO THE LOCATION OF THE OBJECT IN THE SKY WHEN OBSERVED, WITH DIFFERING SUN ANGLES. IT WOULD SEEM THAT A SPIN UP OCCURRED BETWEEN MAY AND SEPT ON THIS MEAGRE DATA.
I HOPE TO OBTAIN FURTHER DATA IN THE NEXT WEEK OR TWO. I CAN SUPPLY THE DATE, TIME, RA AND DEC OF THE OBSERVATIONS. ALONG WITH THE SITE COORDINATES, AND OBSERVED MAGNITUDES, IF SO REQUIRED.

BEST WISHES MAX

Appendix 9
Characteristics of the Ariane 1 Third Stage.

ARIANE 1

1 SPACENET F1

3 Third-stage/SPACENET F1 separation plane
Plan de separation 3eme etage/SPACENET F1

5 Equipment-bay subsystems
Equipements case

7 Vehicle equipment bay (VEB)
Case à équipements

10 Liquid-hydrogen tank
Réservoir d'hydrogène liquide

12 Liquid-oxygen tank
Réservoir d'oxygene liquide

14 Roll and attitude command system
Système de contrôle attitude et roulis (SCAR)

16 Third-stage thrust frame
Bâti moteur du 3eme étage

18 Helium pressurisation tank
Sphere d'helium de pressurisation

20 Interstage 2-3
Jupe inter-étages 2/3

22 Second stage retro rockets (2)
Fusées de freinage 2eme étage(2)

24 N_2O_4 tank
Réservoir N_2O_4

26 Second stage acceleration rockets (4)
Fusées d'accélération 2eme étage(4)

28 1-2 separation plane
Plan de séparation 1/2

30 Interstage 1-2
Jupe inter-étages 1/2

32 First stage retro-rockets (8)
Fusées de freinage 1er étage (8)

34 N_2O_4 tank
Réservoir N_2O_4

36 Intertank skirt
Jupe inter-réservoirs

37 UDMH tank
Réservoir UDMH

39 First-stage torroidal water tank
Tore d'eau du 1er étage

41 Fairings
Carénages

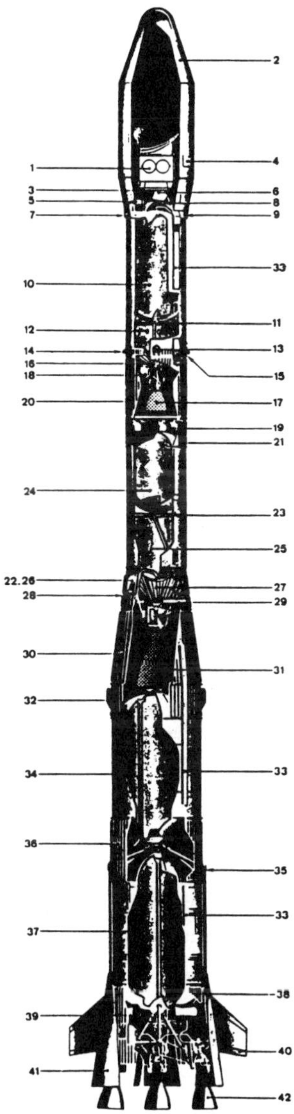

2 Fairing
Coiffe

4 Fairing door
Porte coiffe

6 Payload adaptor
Adaptateur charge utile

8 Third stage sealing membrane
Membrane d'étanchéité 3eme etage

9 VEB antennae
Antennes case

11 Anti-sloshing device
Anti-ballottant

13 Third stage acceleration rockets (4)
Fusées d'accélération 3eme étage (4)

15 2/3 separation plane
Plan de séparation 2/3

17 Third-stage engine HM-7
Moteur HM-7 du 3eme étage

19 Helium pressurisation tanks (3)
Sphères d'hélium de préssurisation (3)

21 Second-stage front skirt
Jupe avant du 2eme étage

23 Anti-sloshing device
Anti-ballottant

25 UDMH tank
Réservoir UDMH

27 Second-stage thrust frame
Bâti moteur du 2eme étage

29 Second-stage torroidal water tank
Tore d'eau du 2eme étage

31 Second-stage engine -Viking IV-
Moteur Viking IV du 2eme étage

33 External cable duct
Gouttiere électrique

35 N_2O_4 pipes (4)
Canalisatioms N_2O_4 (4)

38 First-stage thrust frame
Bâti moteur du 1er étage

40 Fins
Empennages

42 First-stage engines -4 Viking V-
4 Moteurs Viking V du 1er etage

continued...

Appendix 9 (cont'd)
Characteristics of the Ariane 1 Third Stage.

Third Stage (H8)

The third stage has one HM 7 motor, developing in vacuum (chamber pressure: 30 bar) a thrust of 61 kN with specific impulse 440.7 s. The motor pivots on the pitch and yaw axes by means of a gimbal device. Roll control is performed by an ancillary nozzles system supplied with hydrogen gas.

The two tanks holding the liquid hydrogen and oxygen are in aluminium alloy, and have a common intermediate gulkhead (double vacuum loaded skin). The tanks are coated with external thermal protection to avoid overheating of the propellants. The liquid hydrogen and oxygen tanks are pressurized in flight, with hydrogen gas and helium respectively. Burn-out is controlled by the on board computer, using accelerometric data supplied by the inertial guidance platform.

- •Dry mass : 1 115 kg
- •Height : 9.1 m
- •Diameter : 2.6 m
- •Capacity : 8 tonnes of propellants
- •Increment-velocity: 5 576 m/s (F9)
- •Burning time : 560 s (F9)

Vehicle Equipment Bay

Most of the Ariane electrical systems are housed in the Vehicle Equipment Bay (VEB), mounted on the third stage. The VEB comprises a cylindrical shell, forming the base for a conical part, the upper frame of which is used to anchor the payload. The nose fairing rests on the upper frame of the cylindrical part. The conical shell carries the trays on which the VEB equipment is mounted. The on-board computer (OBC) coordinates operation of the various electrical systems of the launcher (sequencing unit, guidance and attitude control, tracking and telemetry systems). Only power and actuating systems are located in the launcher stages.

- •Mass : 272 kg
- •Diameter : 2.6 m
- •Height without adaptor : 1.15 m
- •E band, FM/PCM telemetry (2 203 MHz) : 15 W
- •Two C band radar transponders : 400 W peak

Launcher/Payload Separation

Payload separation is made after third stage burn-out, which is still under attitude control. The payload is released when the clampband is opened by firing a pyrotechnic system. Actual separation is produced by a set of 4 springs arranged on the periphery of the interface frame.

Appendix 10
Debris Size Distributions, March 1, 1987

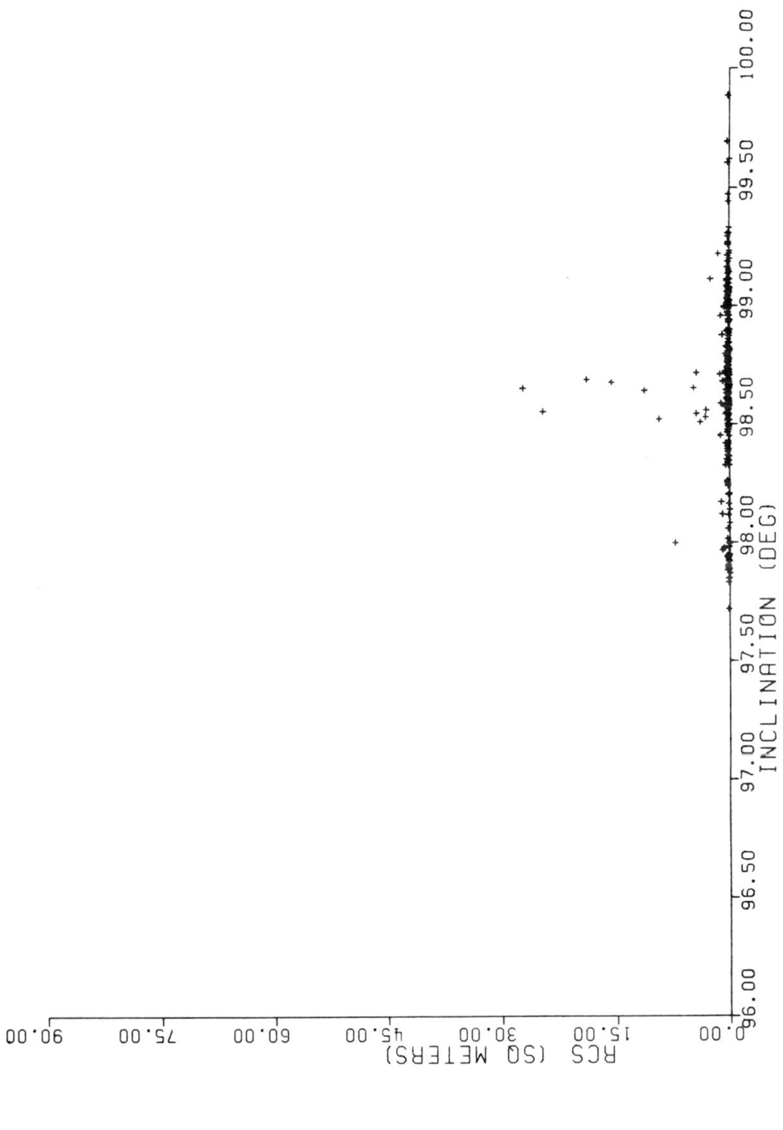

FPS-85 RCS estimates as a function of inclination, March 1, 1987.

continued...

Appendix 10 (cont'd)
Debris Size Distributions, March 1, 1987

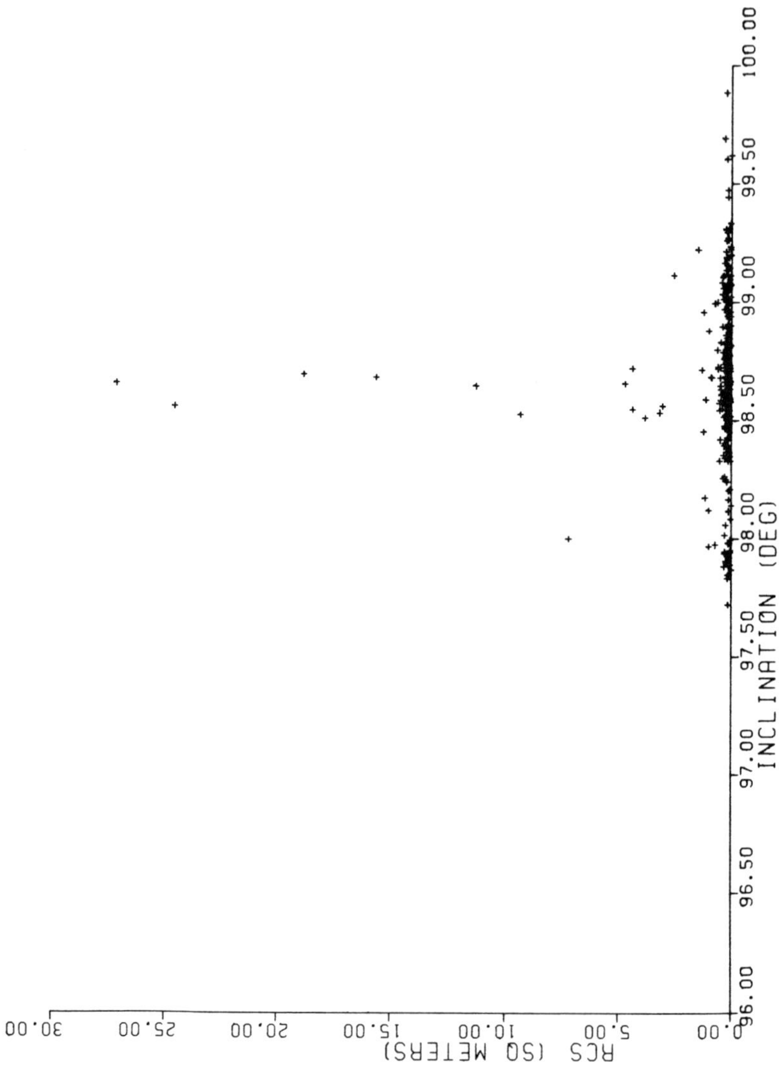

FPS-85 RCS estimates as a function of inclination (expanded view), March 1, 1987.

continued...

Appendix 10 (cont'd)
Debris Size Distributions, March 1, 1987

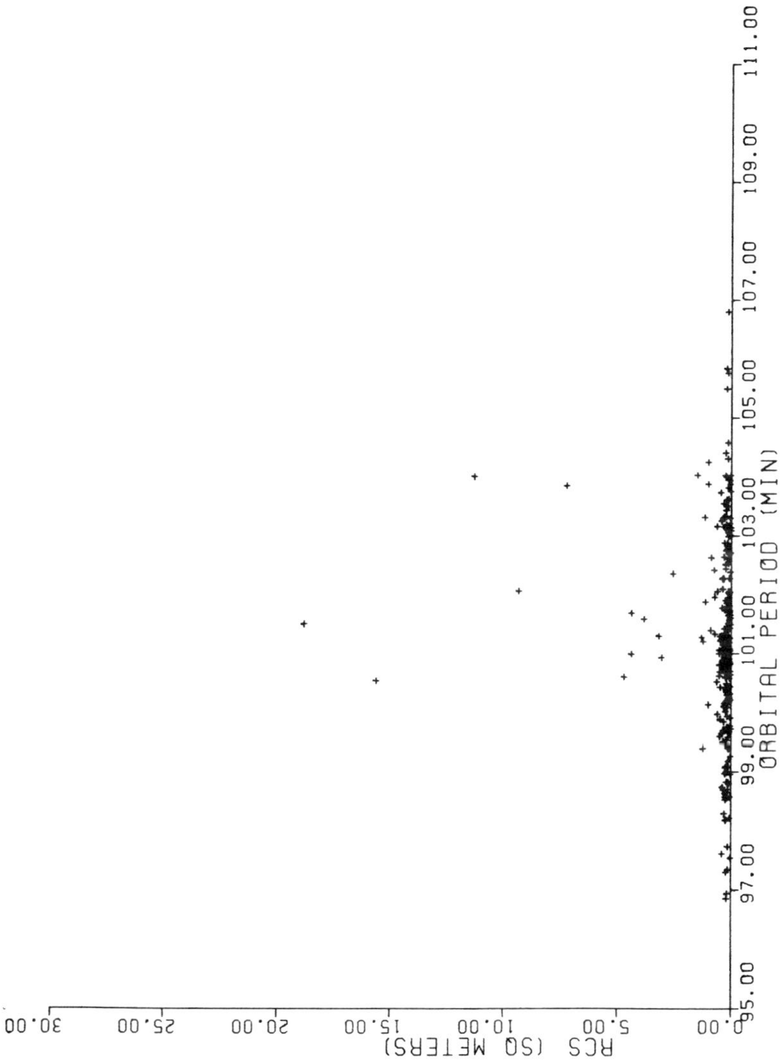

FPS-85 RCS estimates as a function of period, March 1, 1987.

continued...

Appendix 10 (cont'd)
Debris Size Distributions, March 1, 1987

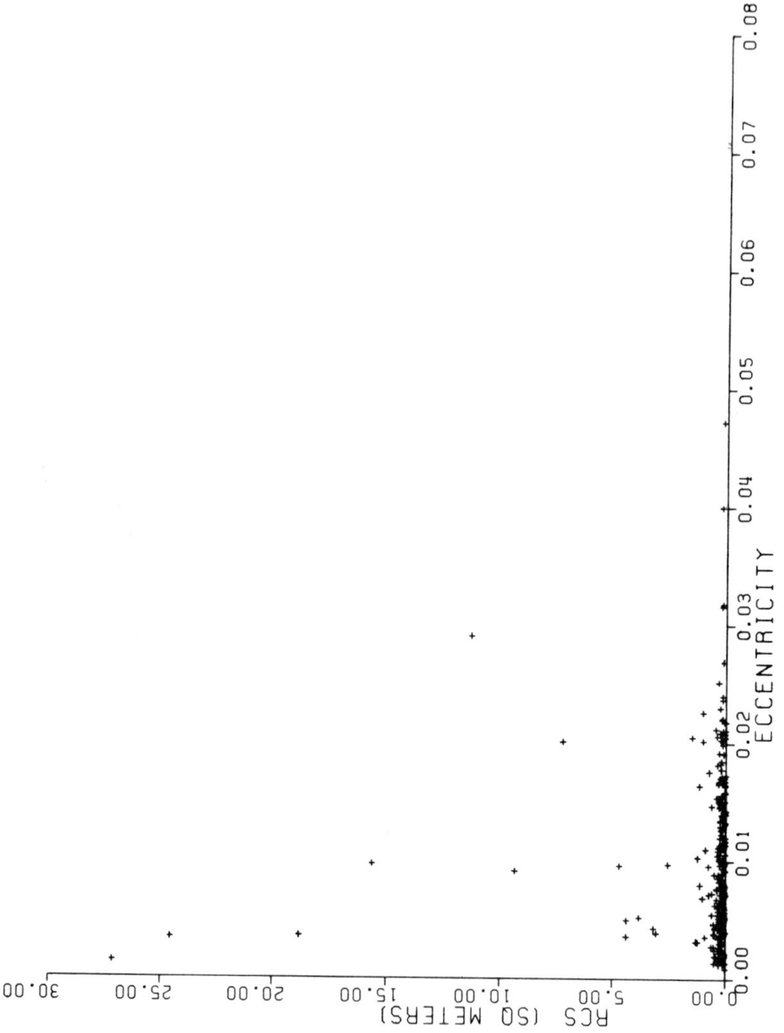

FPS-85 RCS estimates as a function of period (expanded view), March 1, 1987.

continued...

Appendix 10 (cont'd)
Debris Size Distributions, March 1, 1987

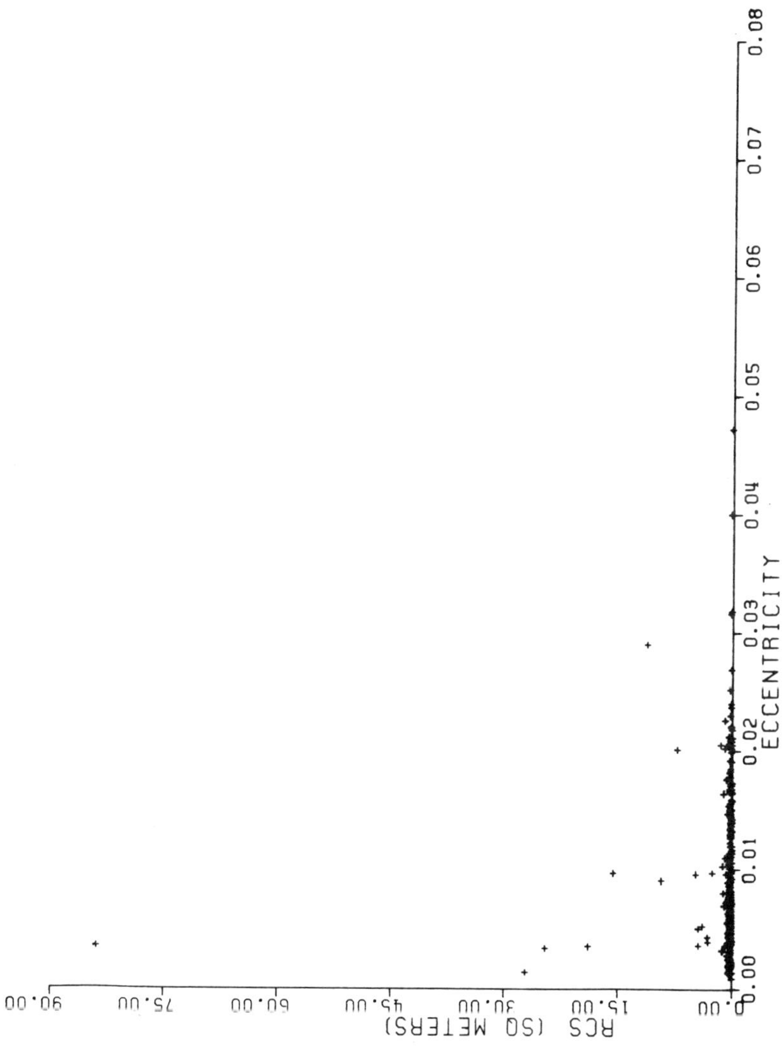

FPS-85 RCS estimates as a function of eccentricity, March 1, 1987.

continued...

Appendix 10 (cont'd)
Debris Size Distributions, March 1, 1987

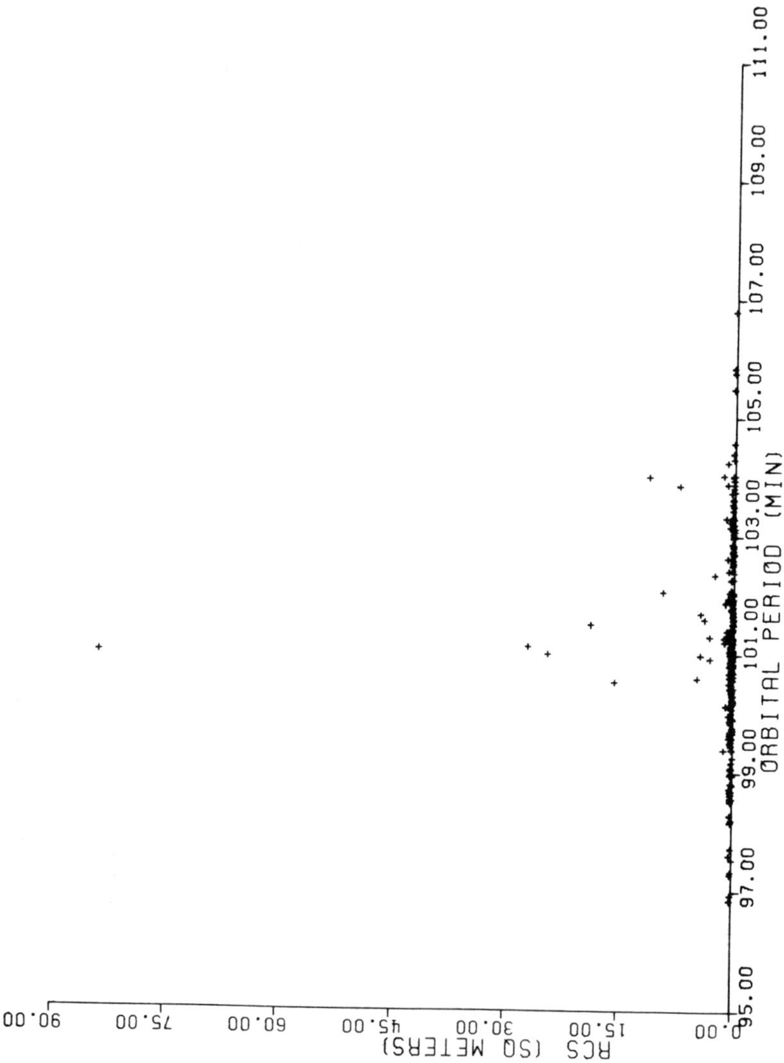

FPS-85 RCS estimates as a function of eccentricity (expanded view), March 1, 1987.

Chapter 5

Explosive Fragmentation of Orbiting Propellant Tanks

F. J. Benz*
NASA White Sands Test Facility, Las Cruces, New Mexico

R. L. Kays†
Lockheed Engineering and Sciences Company, Las Cruces, New Mexico

C. V. Bishop‡
McGean Rohco, Inc., Cleveland, Ohio

M. B. Eck§
Fairchild Space Company, Germantown, Maryland

This chapter examines the explosion properties of orbiting propellant tanks used in the second-stage Delta and third-stage Ariane and vehicle breakups of the Ariane Spot 3 third-stage tanks and the Delta NOAA 3 R/B and Delta Landsat second-stage tanks.

Breakup of propellant tanks can be caused by either physical or chemical explosions of residual propellants, both of which are discussed here. In particular, the explosion properties of propellants used in the U.S. Delta second-stage tanks, which are Aerozine 50 (50% hydrazine and 50% 1,1 dimethylhydrazine) and nitrogen tetroxide (N_2O_4), and those used in the French Ariane third-stage tank, which are liquid hydrogen (LH_2) and liquid oxygen (LO_2), will be presented. In addition, the breakup of the November 13, 1986 Ariane Spot 3 third-stage tank and the breakup of two Delta second-stage tanks (Delta NOAA 3 R/B and Delta Landsat) will be reviewed to investigate the possible role of residual propellants in these breakups. Analysis consisted of review of orbital data, comparison of predicted fragment velocities with the observed fragment velocities, analysis

*Project Director.
†Scientist.
‡Director of Special Projects.
§Technical Staff Specialist.

of the debris patterns, and comparison of the total energy calculated for the debris with the total energy calculated for explosions.

Explosion Properties

An explosion of a tank is a rapid equilibration of a high pressure gas or vapor with the environment where the energy is dissipated in a shock wave, in expansion of the compressed gas or vapor, and in acceleration of the tank fragments.[1] However, in a near vacuum, a shock wave is not generated.

Explosions of a tank can be classified as either physical or chemical.[2] In a physical explosion, the pressure source is generated by nonchemical means, such as by the failure of a tank pressurized with an inert gas (physical pressure burst). If the propellants in the tank are liquids, the explosion process can be augmented by rapid vaporization of the propellants as the tank depressurizes.[3] In contrast, when the pressure source is from rapid chemical reactions, the explosion is referred to as a chemical explosion. Chemical reactions can occur as deflagrations and/or detonations.[4] The effects of any of these explosion processes on the surrounding environment are characterized by the energy release, blast-wave properties, and velocity of the fragments from the tank.[5]

Physical Explosions of Propellant Tanks in Near Vacuum

The energy from a bursting tank is the energy released by the isentropic expansion of the gas or vapor in the tank and is determined by Eq. (1).[6] Typical energy release per unit tank volume (kJ/m^3) of bursting tanks containing gaseous hydrogen (GH_2) and oxygen (GO_2) at various pressures is illustrated in Table 1.

$$E = Pb\ V/\tau - 1\ [1 - (Po/Pb)^{\tau - 1/\tau}] \tag{1}$$

where

E = the blast energy
Pb = the burst pressure
Po = the environmental pressure
V = the volume of contained gas
τ = the specific heat ratio of gas

Table 1 Energy release per unit tank volume from a physical pressure burst of tanks containing GH_2 and GO_2 at various pressures

Pressure (kPa)	Temperature (K)	Energy/Volume (kJ/m3)	
		GH_2	GO_2
400	250	376.5	368.9
300	250	250.2	245.5
200	250	137.2	134.8
100	250	45.7	45.0
50	250	13.9	13.7

In the case of liquid propellants in tanks without a pressurant gas, the energy release from a physical pressure burst is dependent on the propellant vapor pressure and the remaining liquid propellant that undergo rapid vaporization as the tank depressurizes. Typical energy release per unit tank volume for bursting tanks of saturated vapor Aerozine 50 (A-50) and N_2O_4 is illustrated in Table 2 at various temperatures. The total energy release per unit tank volume of A-50 and N_2O_4, when various quantities of the liquid propellant remain in the tanks, is illustrated in Table 3. Equation (1) is used to convert the pressure increase from the flashing liquid propellants to energy release. In this case, the pressure driving the explosion is the sum of the initial vapor pressure and the pressure exerted by the flashing liquid. However, the tank pressure must rapidly decrease before the liquid will flash, implying that the total energy may be released in steps (namely, energy released from initial vapor pressure and then energy released from the flashing liquid propellants).

When the tanks are pressurized with an inert gas, as is the case of the A-50 and N_2O_4 tanks, the energy release from the pressurant gas must also be included. Table 4 gives total energy release per unit tank volume at various GN_2 pressures.

Chemical Explosions of Propellant Tanks

Propellants can chemically explode by undergoing deflagrations and detonations. A deflagration is defined as a burning process in which the flame travels at a subsonic velocity relative to the unburned materials.[7] A

Table 2 Energy release per unit tank volume from physical pressure bursts of tanks with saturated vapors of A-50 and N_2O_4 at various temperatures

Temperature (K)	Pressure (kPa)		Energy/Volume (kJ/m3)	
	A-50	N_2O_4	A-50	N_2O_4
270	5.6	30.1	0.5	7.0
280	8.6	50.3	0.6	17.7
290	13.2	80.1	1.5	42.1
300	19.9	132.4	3.4	94.2

Table 3 Energy release per unit volume from a bursting tank with various quantities of liquid A-50 and N_2O_4 remaining in the tank (290 K)

Volume Liquid in Tank, %	Energy/Volume, kJ/m	
	A-50	N_2O_4
5.0	300.1	290.3
2.0	83.1	130.4
1.0	27.9	82.8
0.5	12.2	60.2
0.1	2.7	45.1
0.0	1.5	42.2

detonation is defined as a burning process in which the chemical reaction generates and takes place inside a shock wave that travels at supersonic velocities relative to the unburned materials.[4]

Mixtures of GH_2 and GO_2 will deflagrate and detonate when exposed to an ignition source.[4] When LH_2 and LO_2 are mixed in a closed container, they form high-pressure multiphase mixtures consisting of GH_2, LH_2, and solid oxygen.[8] These mixtures will undergo deflagrations,[9,10] but whether they will detonate is still being investigated.[11]

Vapor or liquid A-50 and N_2O_4 will ignite and burn upon contact.[12,13] Their deflagration properties are controlled by the mixing process. However, it is unlikely that a mixture of A-50 and N_2O_4 will detonate because they will react before a mixture can be established to support a stable detonation. Vapor A-50 will deflagrate by itself (without an oxidizer), but the deflagration appears to occur only at high pressures and temperatures (above 207 kPa and 372 K) whereas liquid A-50 does not appear to ignite and burn at these same conditions.[14] Hydrazine, one of the constituents of A-50, will deflagrate and detonate as vapor in the absence of an oxidizer.[15]

Pressure Increase

The energy generated from a chemical reaction is the difference between the enthalpy of formation Hf of the products and the reactants. The pressure increases depend on the increase in temperature and the change in moles. The pressure increase ratio from a deflagration (constant-volume combustion) can be estimated by determining the adiabatic flame temperatures and using the following relation:

$$\mathrm{Pcv}/Pi = (\mathrm{Tad}/Ti)(nf/ni) \tag{2}$$

Table 4 Energy release per tank volume from physical pressure bursts of tanks with A-50 or N_2O_4 pressurized with GN_2

GN_2 Pressure, kPa	Energy-release/Volume, kJ/m^3	
	A-50	N_2O_4
A. 5% volume of tank containing liquid propellant.		
1400	2500	2540
1000	1810	1850
600	1170	1200
200	560	593
0	300	290
B. No liquid propellant in tank (vapor only).		
1400	1880	2070
1000	1250	1400
600	650	780
200	145	240
0	1.5	42

where

Pcv = constant-volume-combustion overpressure
Pi = initial pressure
Tad = adiabatic flame temperature
Ti = initial temperature
ni = initial number of moles
nf = final number of moles

In the case of H_2 and O_2, the pressure increase is caused solely by the temperature increase because $nf/ni = 0.67$ whereas, in the case of A-50 and N_2O_4, $nf/ni = 2.7$. The pressure-increase ratio for detonations Pd/Pi is determined from the Chapman-Jouguet (C-J) properties for the reacting materials and is obtained from various codes, such as the Gordon and McBride (1976) code.[16] The maximum Pcv/Pi for deflagrations and the maximum Pd/Pi for detonations of materials normally occur at stoichiometric concentrations and decrease as the mixture is made rich or lean in fuel. Typical Pcv/Pi for deflagrations [Eq. (2)] and Pd/Pi for detonations of stoichiometric mixtures of GH_2 and GO_2 and vapor A-50 and N_2O_4 are provided in Table 5.[15] The Pd/Pi for detonation of A-50 and N_2O_4 are included for comparison, even though it is believed that this mixture cannot be established to support stable detonations.

Detonation pressures (Table 5) are the pressures in the detonation wave that travel at enormous speeds. For example, the detonation velocity of GH_2 and GO_2 is approximately 3000 m/s. The pressure in front of a propagating detonation wave is the initial ambient pressure, whereas the pressure behind the detonation wave drops quickly to near the constant-volume combustion pressure. When a detonation wave impacts perpendicular to a flat surface and reflects, the pressure acting on the surface can more than double[6]; $Pm/Pi > 46$, as in the case of GH_2 and GO_2. Even though large, these detonation pressures are transient conditions, and the pressure responsible for accelerating fragments in a bursting tank will be the lower constant-volume combustion pressure. However, the Pd's are important because they can provide the transient conditions that can cause tanks to fail and can influence the type of tank failure.

If a deflagration undergoes transition to detonation (DDT) in a tank, much larger detonation pressures can be achieved. The deflagration increases the total pressure in the tank, and DDT will occur at a higher initial pressure. If a deflagration of GH_2 and GO_2 raises the pressure in the tank by a factor of 5 before DDT, the detonation pressure ratio becomes $Pd/5\,Pi = 23$ or $Pd = 115\,Pi$.

Table 5 Pressure-increase ratios for deflagration and detonation of stoichiometric mixtures of GH_2 and GO_2 and A-50 and N_2O_4 vapor

Mixture	Initial Temp., K	Pcv/Pi	Pd/Pi
GH_2/GO_2	250	9/1	23/1
A-50/N_2O_4	300	34/1	72/1

The rate of pressure increase dP/dt is especially important in deflagrations because, if the deflagration is slow compared to the failure of the tank, the bursting tank may not be driven by the maximum pressures listed in Table 5. The enormous speed of a detonation causes the maximum pressure to be established before the tank has a chance to fail and relieve pressure. The dP/dt for a deflagration is strongly dependent on the burning velocity, as shown in the following for central ignition in a spherical tank[17]:

$$dP/dt = k\,Pi\,Ub^3 t/V \tag{3}$$

where

k = dimensional constant
t = time
Ub = burning velocity
V = volume of tank

For example, burning velocity for GH_2 and GO_2 can vary from a few meters per second for laminar burning[7] to several hundred meters per second for turbulent burning.[18]

Energy Release

The energy release when a propellant tank bursts from an internal chemical explosion can be determined from Eq. (1), assuming a constant-volume combustion pressure increase. Table 6 shows energy release per unit tank volume of a bursting tank for stoichiometric mixtures of GH_2 and GO_2 and vapor A-50 and N_2O_4 at various initial pressures. The highest initial pressure for the A-50 and N_2O_4 mixture will be where a saturated vapor of A-50 at 300 K can produce a stoichiometric mixture (20 kPa for A-50 and 20 kPa

Table 6 Energy release per unit tank volume of a bursting tank from combustion of stoichiometric mixtures of GH_2 and GO_2 and A-50 and N_2O_4 vapor at various initial pressures

Pi, kPa	*Ti*, K	Pcv, kPa	Energy/Volume, kJ/m^3
A. GH_2 and GO_2 mixtures.			
400	250	3600	6454
200	250	1800	2827
100	250	900	1187
50	250	450	470
10	250	90	40
B. Vapor A-50 and N_2O_4 mixtures.			
40	300	1360	1943
20	300	680	800
10	300	340	305
5	300	170	110
1	300	34	7

for N_2O_4). In reality, the A-50/N_2O_4 tanks will be pressurized with GN_2 and will have residual liquid propellants.

For example, consider propellant tanks containing 2% tank volume of liquid A-50 and sufficient liquid N_2O_4 to produce stoichiometric burning, and both tanks pressurized to 1378 kPa with GN_2. The total energy release per unit tank volume of a bursting tank from total combustion of all propellants and release of the pressurized GN_2 will be 11,630 kJ/m^3.

Explosion Energy from Known Fragment Velocities

The fragmentation energy can be calculated by using the change in orbital parameters and by assuming that most of the debris is observed by radar. To approximate the fragmentation energy by this method, the initial mass of the object must be known, and the ratio of propellant mass to inert mass is assumed to be negligible. Also, the orbital parameters must be known with great precision. The total energy of a satellite is given by Eq. (4), and the fragmentation energy may be estimated by Eq. (5):

$$E/m = GM/2ra \tag{4}$$

$$(E/m)\text{frag} = \text{abs}[\overline{E/m(f)} - \overline{E/m(i)}] \tag{5}$$

where

E/m = total energy per mass of orbiting material
GM = product of the gravitational constant and the mass of the Earth
ra = semimajor axis of the orbit
$\overline{E/m(f)}$ = average final E/m of fragment
$\overline{E/m(i)}$ = average initial E/m of debris

Response of Tank Fragments

The energy release from a bursting tank manifests itself in the acceleration of the fragments from the failed tank and expansion of the pressurized gas or vapor. The portion of the total energy that actually goes into accelerating the fragments will vary with the failure mechanism of the tank.

Knowing the initial velocities of fragments allows deriving the initial force on them and thus the burst pressure. From that can be estimated the burst energy of the explosion process, physical or chemical.

The velocity of the fragments from a bursting tank is determined from the momentum change Δmv where the initial velocities of fragments are zero relative to the motions of the tank, as expressed in the following equations. (The drag coefficient in space approaches zero, and the initial velocities of the fragments are maintained for long periods of time.)

$$m\Delta v = \Delta F\,\Delta t \tag{6}$$

where $\Delta F = \Delta P\,A$ and for a tank of constant thickness $m = PA$.

Therefore

$$\Delta v = \Delta P \, \Delta t / l \tag{7}$$

where
A = area of the fragment
l = thickness of the fragment
m = mass of the fragment
ΔP = initial minus final pressure
Δv = initial minus final velocity of fragment
Δt = time P is acting on fragment
ρ = density of fragment

The difficulty involved in solving Eq. (7) is knowing t, which is dependent on how the tank fails. Operators such as the Pisces code[19] allow estimates of t based on the time required to vent pressure as the tank fails. However, t also depends on the failure mechanism of the tank. The Pisces code is capable of modeling various types of failure mechanism, such as the two-dimensional Clam model (Fig. 1a), Half-Segment model (Fig. 1b), and Octant model (Fig. 1c). The code is also capable of three-dimensional analysis but requires more computation time.

Also illustrated in Fig. 1 is the velocity distribution possible depending on the type of failure. The Clam model predicts a large distribution of fragment velocities whereas the Octant model assumes a tank failed uniformly and predicts that all fragments leave at the same velocity.

Baker et al.[1] developed models and computer codes for analyzing bursting tanks to obtain estimates of initial fragment velocities. In their model, the energy of the confined gas or vapor is partitioned between the kinetic energy of the fragments, the energy of the gas escaping between the cracks of the fragments as they form, and the energy of the expansion of the internal gas. The vessel is assumed to break instantaneously into n fragments and, in the case of cylinders, into n strips that move radially from the center of the cylinder. Based on information from their codes, figures were generated and published in workbook fashion that easily allow quick estimates of initial velocities once the mass ratio (mass of gas/mass of tank) and pressure ratio (burst pressure/ambient pressure) are known. This method assumes all fragments leave at the same velocity (similar to the PISCES and Octant model). This model can not determine fragment-velocity distributions.

Fragment velocities can also be estimated using the Gurney Analysis,[5] which was originally developed for fragmenting munitions, in the following form:

$$vf = (-2\Delta e)^{0.5}\{(C/M)/[1 + (0.5)(C/M)]\}^{0.5} \tag{8}$$

where
C = mass of explosive or pressurized gas
M = mass of total fragments or tank
Δe = energy per unit mass
vf = fragment velocity

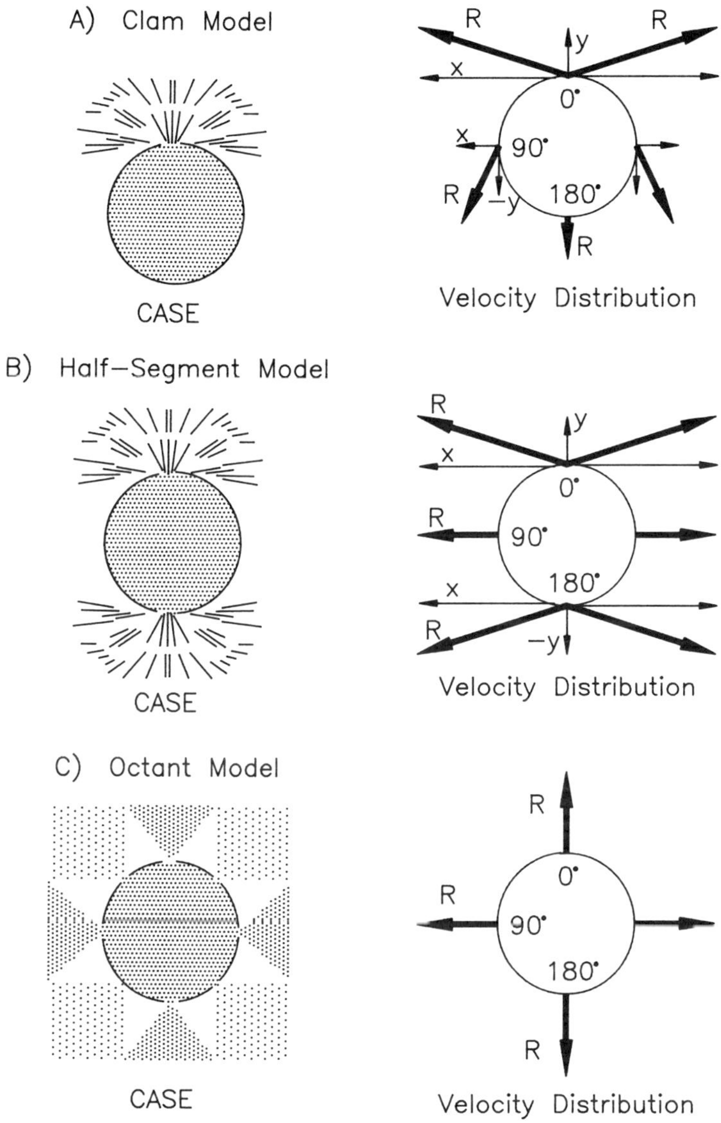

Fig. 1 Types of tank failures and corresponding fragment velocity-vectors.

Table 7 Methods for estimating fragment velocities from a physical pressure burst of a tank in near vacuum pressurized to 1378 kPa with GN_2

Method	Fragment Velocity, m/s
Pisces Code (Clam model)	20–385
Baker et al model	340
Gurney analysis	255

This analysis assumes that the kinetic energy of the fragments plus the energy of the expanding gas or vapor move out at more or less the same initial velocity, and that the velocity of the gas or vapor increases from zero to that of the initial velocity of fragments. The energy/mass term Δe can be obtained from Eq. (1), and the mass is that of the pressurized gas or vapor. This method, like the Baker model, can not determine fragment-velocity distributions; but, as for the Baker model, the fragment velocities can be quickly estimated.

Table 7 compares the fragment velocities estimated from the three techniques for a tank made of 410 stainless steel that is 1.39 m in diam and 0.002 m thick, with a volume of 4 m. The tank is pressurized to 1378 kPa with GH_2 at ambient temperature, and subsequently undergoes a physical pressure burst.

Recognizing Type of Tank Failures

Nonpressurized orbital components that collide with a solid will fragment, based on the mechanism proposed by Su and Kessler (1985)[20] and will produce a debris pattern, as described by Kling (1986).[21]

Pressurized and failing at its burst pressure, a tank will typically tear longitudinally (Pisces Clam model) and fracture based upon the constructive interference of tensile waves in the tank. Usually the longitudinal tears will be in halves, in four pieces, or in eight pieces. As the tank opens, centripetal forces will straighten the curvature, (which Nance calls wet noodling)[22] and will eject fragments tangentially, the debris pattern resembling diverging rays from the tank axis, as shown in Fig. 2a. In the recent Titan 34-D accident, a similar debris pattern was observed for the solid-propellant rocket motor that was disabled (range destruct) by a linear shaped-charge.[22] In this type of failure, only the smallest pieces will achieve the highest velocities. This is illustrated in Fig. 2a by the change in the velocity vectors between the 0 and 90 deg. Only small pieces that break near the 0-deg zone will attain the highest velocities because large pieces that break in this region will have a larger portion of their area affected by zones where the velocities vectors are smaller. However, the trend breaks down for tanks where the mass-to-surface areas are no longer constant, such as tanks that vary in thickness. It is also possible for the tank to fail as the half-segment model (Fig. 2b).

Acted on externally, as by a micrometeor striking and causing a tear on one side, the tank would still fail in the manner illustrated in Fig. 2a. However, the velocities of the debris would be lower because of the lower tank pressures.

Tanks undergoing internal detonations of propellants can produce random debris patterns because the pressure increase (constant-volume combustion) in the tank is achieved almost instantaneously and can reach pressures much higher than the burst pressure of the tank. For example, detonation of GH_2 and GO_2 at initial pressure near the tank burst pressure can produce pressures nine times the burst pressure of the tank. In this case, the entire tank would tend to fail in all directions, with many fragments having a small velocity distribution, similar to the Pisces/Octant model illustrated in Fig. 2c.

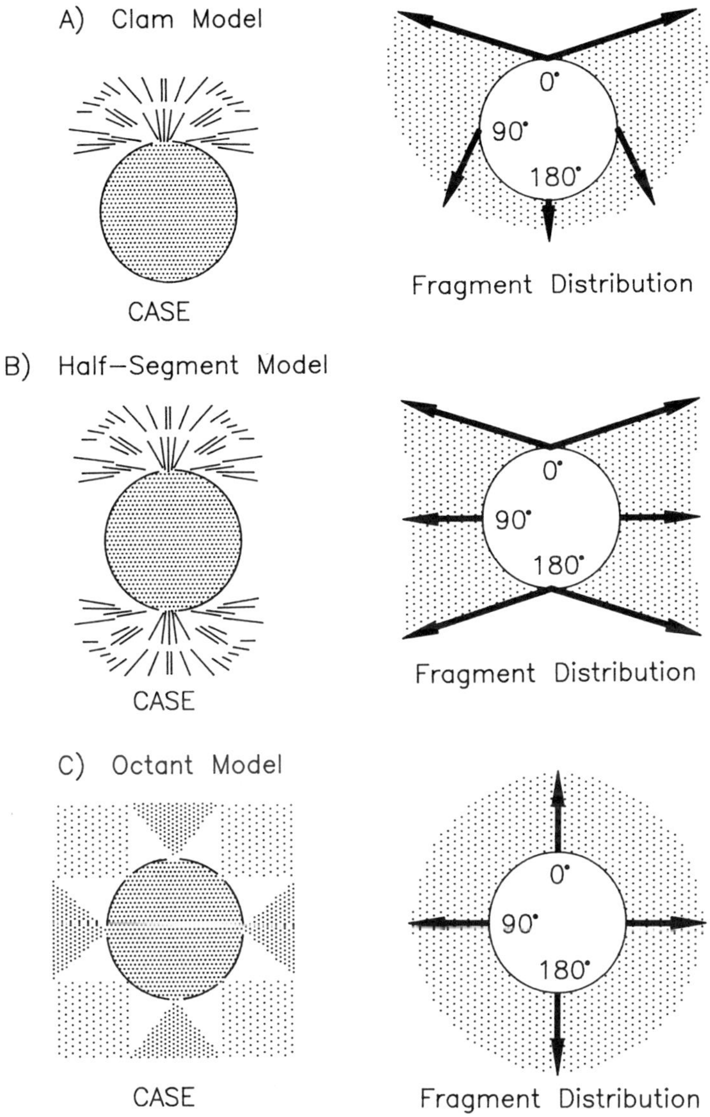

Fig. 2 Fragment distribution for types of tank failure.

These observations lead to the picture in Table 8.

Ariane and Delta Breakups

Debris data for the Ariane Spot 3, Delta NOAA 3 R/B, and Delta Landsat were supplied by the Space Sciences Branch, Solar System Exploration Div., NASA Johnson Space Center; this information is included in the Appendix. Information on the Ariane third-stage tank and breakup

Table 8 Recognizing pressure failure from velocity and distribution of debris

Pressure at burst implies burst energy	Velocity distribution of debris	Geometric distribution of debris	Cause of tank failure
$P \gg$ burst strength	Small	Random	Detonation
$P \approx$ burst strength	Large	Two rays	Deflagration or physical pressure burst
$P <$ burst strength	Large	Two rays	Fatigue or micrometeor penetrated one wall

conditions comes from the *Ariane User's Manual*[23] and from Hergott.[24] Information on the Delta second-stage tanks was obtained from a McDonnell Douglas Astronautics Company report.[25] Information specifically on the Delta Landsat and NOAA 3 R/B was obtained from Gray.[26]

Fragment velocities were determined using the Pisces Code, Baker et al. model, and Gurney analysis and then compared with the reported velocities for the Delta NOAA, Delta Landsat, and Ariane Spot 3. The orbital-debris data were graphically prepared to determine similarities and differences in the three breakup events. The explosion energies for the various breakups were then compared using Eqs. (1), (2), (5), and (8). Finally, orbital data for the Ariane Spot 3 were used to determine the spread pattern of the debris.

Predicted and Reported Fragment-Velocities

Velocities were predicted for cases in which the propellant tanks underwent only a physical pressure burst (no chemical reaction) and an internal chemical explosion. The initial conditions used for the Ariane were estimated as the conditions of the tank at the time of the breakup. The conditions used for the Delta NOAA and Delta Landsat were conditions of the tank just after mission completion. The predicted velocities were then compared with the reported velocities to illustrate the most likely cause for tank breakup.

Ariane Spot 3 Third-Stage Tanks

Table 9 compares results from several techniques for predicting the fragment velocity, where the breakup of a propellant tank was assumed to be caused by a physical pressure burst or detonation of GH_2 and GO_2. The conditions evaluated for the physical pressure burst case were at the design burst pressures of the tanks (370 kPa for the H_2 tank and 340 kPa for the O_2 tank). The detonation case was evaluated based on the quantity of GH_2 and GO_2 thought to be left in the tank (7 kg of H_2 and 22 kg of O_2). When these results are compared with the reported velocities, it is evident that the detonation case at these conditions does not agree with the reported velocities.

Detonation or rapid deflagration of H_2 and O_2 could, under the right conditions, cause the reported velocities; but for this to occur, both tanks would have to leak in such a way that the pressure in the O_2 tank would

Table 9 Predicted fragment velocities for the Ariane third-stage tank breakup compared with the reported fragment velocities of 15–278 m/s

Method	Type of burst	Predicted velocity m/s	
		H_2 Tank	O_2 Tank
Pisces (Clam)	Physical pressure	28–208	2–262
Pisces (Half-Segment)	Physical pressure	46–182	80–218
Pisces (Octant)	Physical pressure	105	137
Gurney	Physical pressure	188	169
Pisces (Octant)	Detonation[a]	500	500
Gurney	Detonation[a]	570	570

[a]Both tanks taken as one tank in the analysis.

never exceed the 50-kPa pressure differential until the O_2 tank attained a pressure of 75 kPa. The H_2 would have to continue to leak until it achieved a pressure of 25 kPa, thereby causing the bulkhead failure and allowing the mixing of the H_2 and O_2. At these conditions the detonation or rapid deflagration would produce the proper overpressure to fail the tank and produce the reported fragment velocities. Even before this, however, the mixture would have to be exposed to an ignition source. These series of events, though not impossible, seem unlikely because, if any one event occurred out of order, the overpressures required to produce the reported velocities would not be attained. In the case of the physical pressure release, any one of the following events by itself could lead to the explosion event: fatigue of the tank, collision of the tank with a small particle (space junk), failure of any of the two release valves, or failure of the bulkhead that caused the tanks to fail.

Even if the pressure in either of the propellant tanks were at or slightly below their relief valve settings (315 kPa for the H_2 tank and 235 for the O_2 tank), the predicted velocities would still be in agreement with the reported velocities for a physical pressure burst. For example, the predicted fragment velocities at the relief pressure were 95 and 120 m/s for the H_2 and O_2 tanks, respectively, for the Pisces (Octant) model, and 170 and 155 m/s for the H_2 and O_2 tanks, respectively, for the Gurney analysis.

These results strongly suggest that the Ariane third-stage tank breakup was caused by a physical pressure burst that was initiated by overpressurization of the tank or by an external source. An internal chemical explosion of the tank appears to be highly unlikely as the cause of this breakup.

Velocities of Fragment from Delta NOAA and Delta Landsat Second-Stage Tanks

The initial conditions of the Delta NOAA and Delta Landsat shortly after orbital deployment and the reported fragment velocities are given in Table 10. The Delta NOAA and Delta Landsat second-stage tanks were pressurized to approximately 1380 kPa with helium. The design burst pressures of the tanks are approximately 2750 kPa. It was assumed for this evaluation that the tanks failed at the design burst pressures from internal pressuriza-

Table 10 Reported fragment velocities from the Delta NOAA and the Delta Landsat second-stage tank breakup

Event	Initial conditions at deployment	Reported fragment velocities, m/s
Delta NOAA, (of liquid propellants 133 kg)	A-50 tank: 896 kpa N_2O_4 tank: 792 kPa GHe tank: 689 kPa	8–485
Delta Landsat, (of liquid propellants 88 kg)	A-50 Tank: 1275 kPa N_2O_4 Tank: 1378 kPa GHe Tank: 4134 kPa	10–400

tion caused by the isothermal reaction of A-50 and N_2O_4. Predicted fragment velocities are given in Table 11 using the following methods: Pisces Code (Clam model), Baker et al model, and Gurney analysis. The thickness of the second-stage tanks vary from 0.0020 m (cylindrical portion of the tanks) to 0.0011 (domes), and thus require a three-dimensional analysis. Time or resources did not permit this; but a two-dimensional analysis was performed using the Pisces Code (Clam model) for tanks with constant thicknesses of 0.0011 and 0.002 m. This approach allowed an easy way to obtain the magnitude of velocities that would be expected from tanks of varying thickness. This same approach was used with the Baker et al. model and the Gurney analysis.

The predicted velocities compared reasonably well with cases of the propellant tanks bursting at design pressure. This also implies that a leak in the bulkhead separating the two propellants caused an internal overpressurization of the tanks. The reported fragment velocities for Delta Landsat were lower than those of the Delta NOAA, perhaps because burst pressure of the Delta Landsat was lower from the start, or the tank fatigued with time. In addition, an external source (such as space junk or linear-shape destruct charge) could have caused the initial failure of the tank, which then burst at a lower internal pressure.

Similarities and Dissimilarities of the Debris Data

Three-dimensional illustrations of the debris change in velocity, inclination, period, and radar cross section (RCS) for the three tank breakups are given in Figs. 3–5. In all three tank breakups, the largest change in velocity, inclination, and period were for debris with the smallest RCS. The debris with the largest RCS showed the smaller change in velocity, inclination, and period, with the exception of the Delta NOAA, in which several fragments of large RCS exhibited large velocities and changes in inclination. The Delta NOAA and the Ariane produced debris with large changes in inclination and period; the period mostly increased. The change in inclination of the debris for the Delta Landsat was much less, and the debris exhibited both increasing and decreasing periods. Especially interesting was the debris of the Delta Landsat, which decreased in period but changed little in inclination.

These data indicate that the tank breakups were not geometrically uniform, as predicted by the Octant model, and that the debris were produced by explosive events similar to those predicted by the Clam or Half-Segment models. In addition, the Landsat breakup shows signs that it produced fragments that were directed tangential to the original orbit of the propellant tank, as indicated by debris with decreasing periods and inclinations close to the original inclination of the tanks. Thus, these debris data support

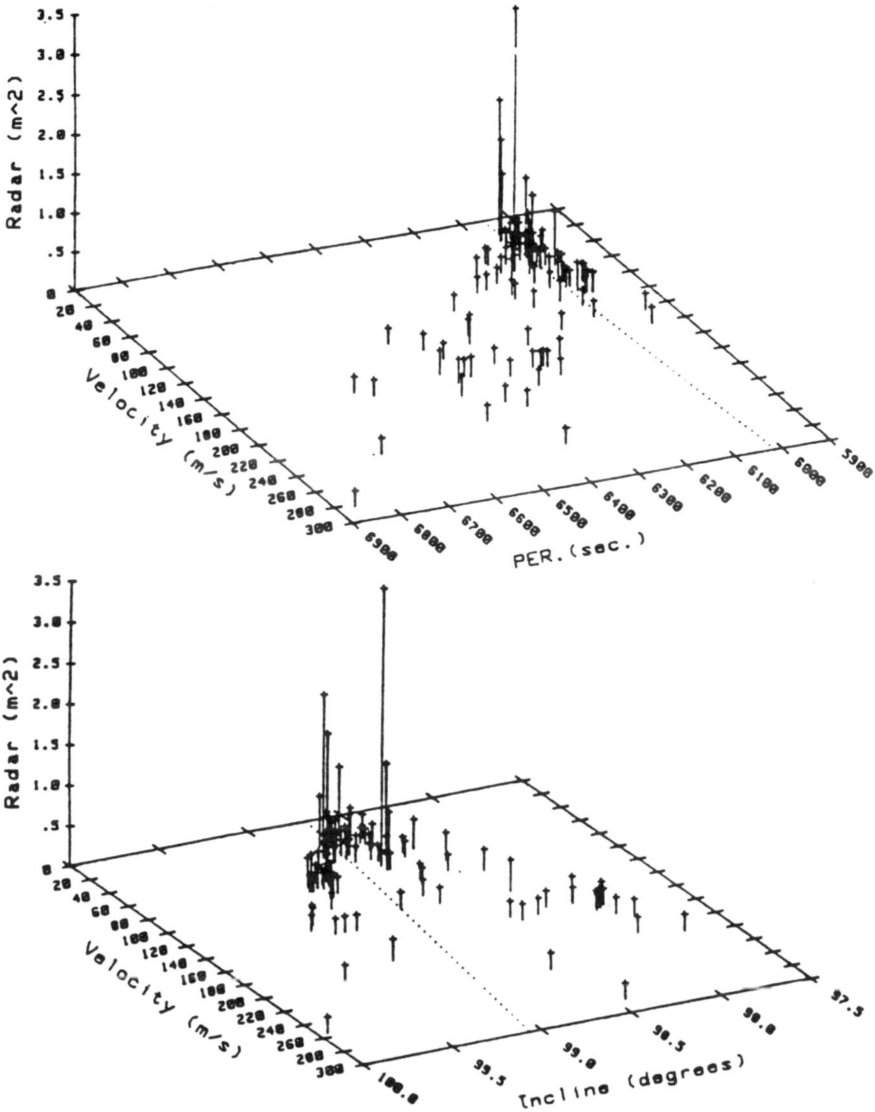

Fig. 3 Orbital-debris data for the Ariane third-stage tank.

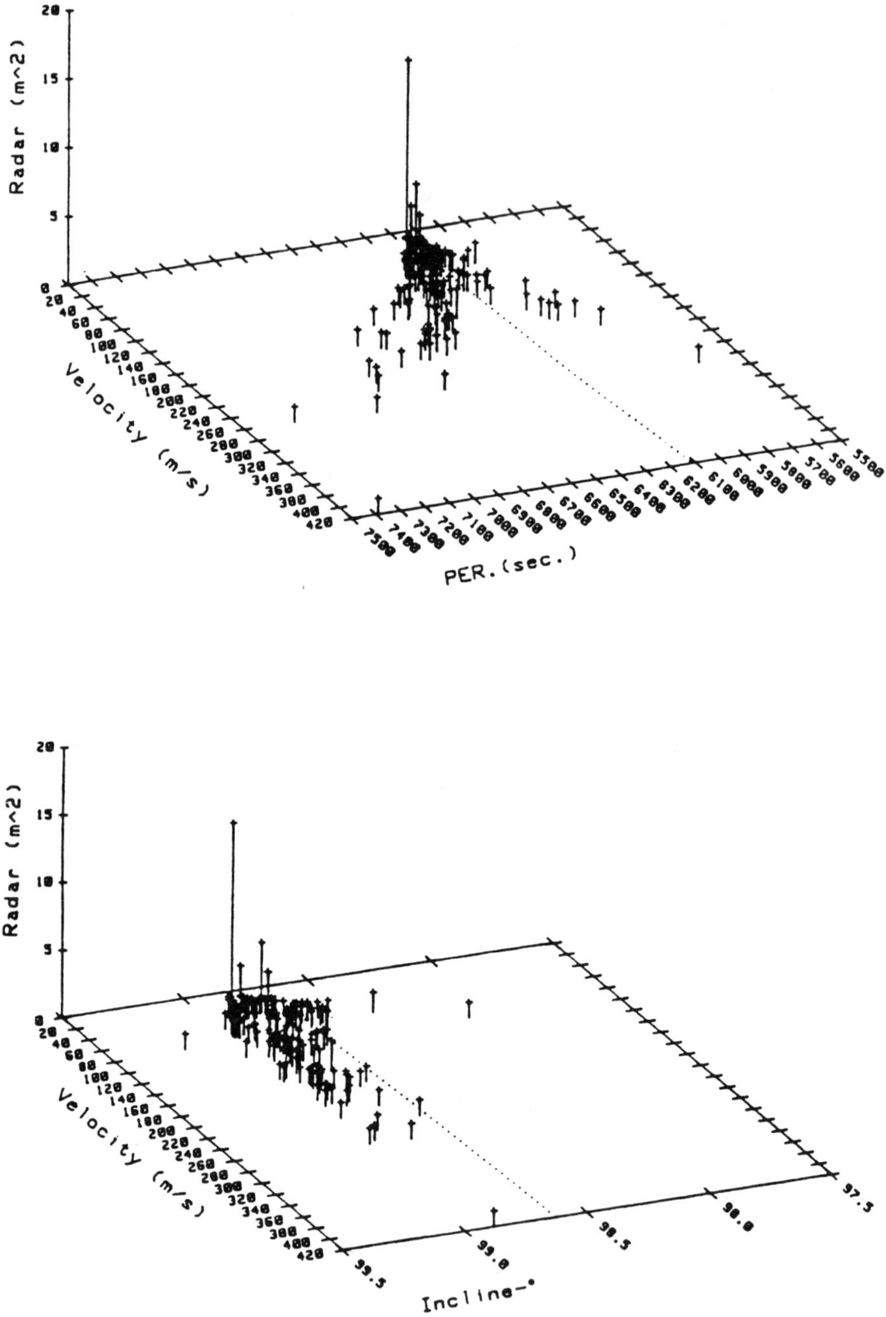

Fig. 4 Orbital-debris data for the Delta Landsat second-stage tank.

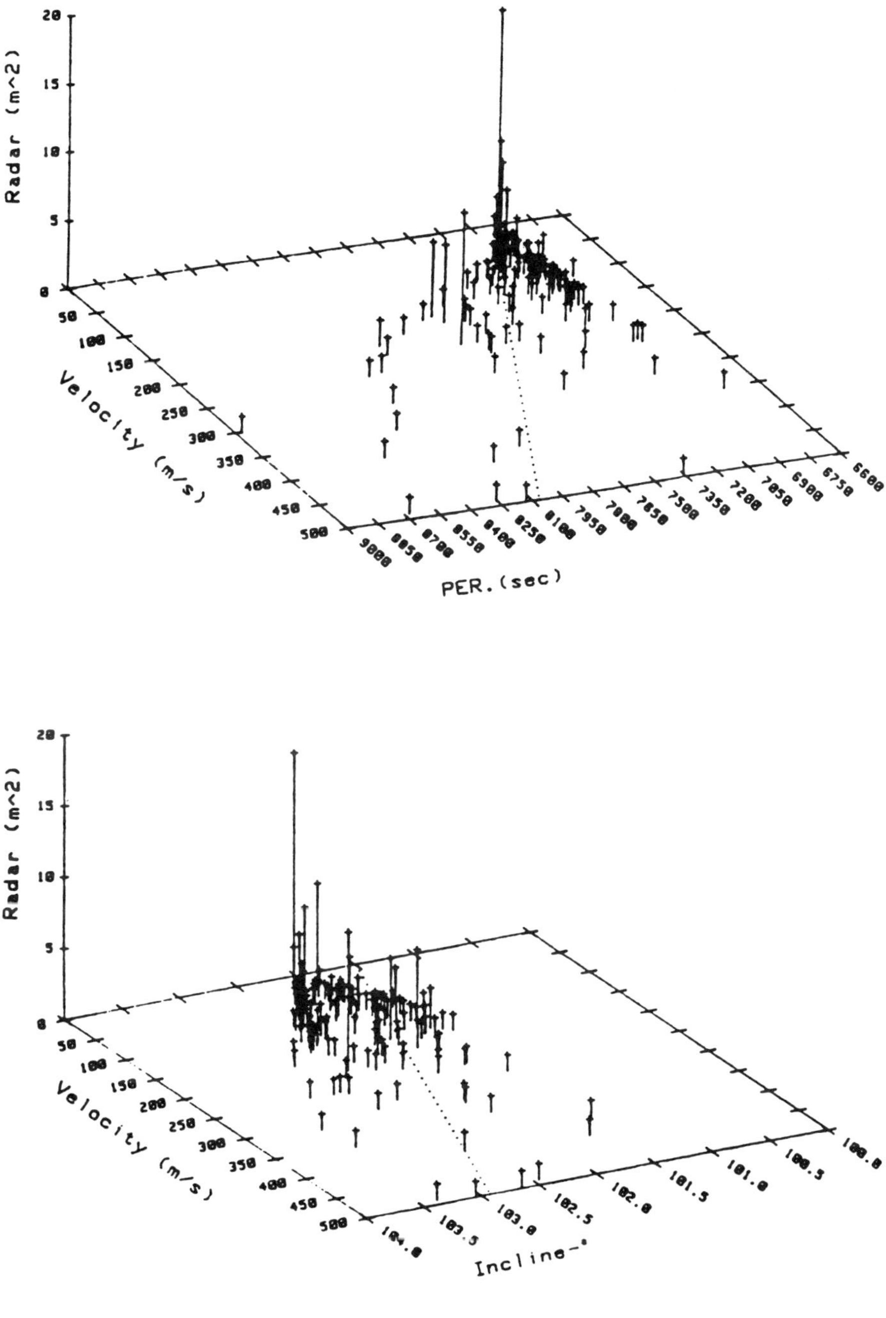

Fig. 5 Orbital-debris data for the Delta NOAA second-stage tank.

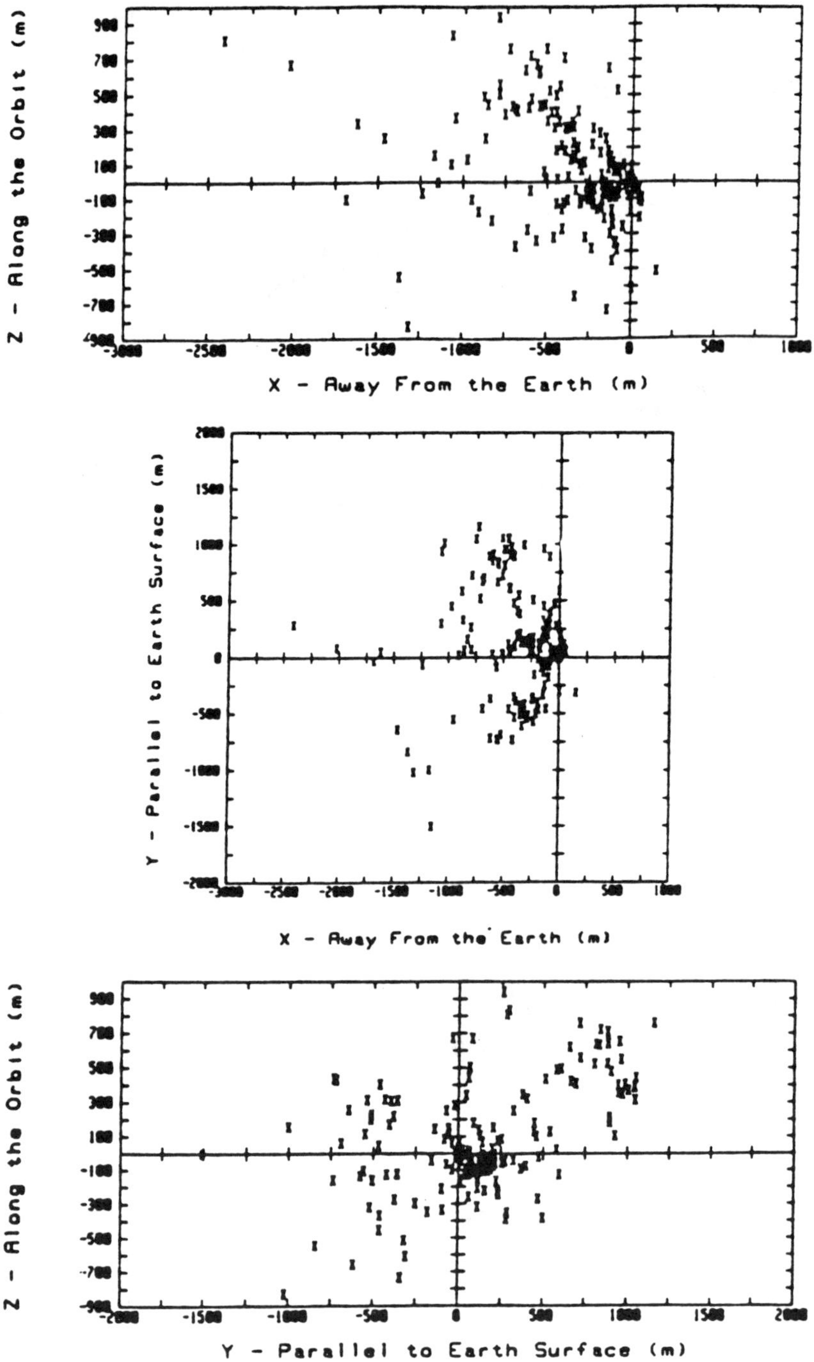

Fig. 6 Orbital-debris pattern in the z – x, y – x and z – y planes for the Ariane third stage.

the conclusions in the previous section that the breakups of the tanks were caused by physical pressure bursts and that they failed as predicted by the Clam model.

Explosion Energy by Orbital Parameters for Ariane Third Stage

The Ariane data indicate that the clamp band (16616) adjacent to the orbiting third-stage tank was unaffected by the explosion. The E/m of this piece was calculated as 27.50 MJ/kg, using Eq. (4). This is assumed to be the average initial E/m for all debris. Calculating the absolute individual final E/m for each piece of debris and averaging the result yields an average final E/m of 27.53 MJ/kg, 0.1% greater than the initial E/m. Because the exact mass of the third-stage tanks and motor are unknown, a mass of 1000 kg was assumed (560-kg tanks and 440-kg motor). The total fragmentation energy of the debris was then calculated as roughly 30 MJ. However, this does not represent the total energy of the explosion. Results from the Pisces Code predict that approximately 30% of the total explosion energy was imparted to the fragment, and the remaining energy was used in the expansion of the gas or vapor. Therefore, the total energy determined from this method should be approximately 100 MJ.

The energy calculated from Eq. (1) at the conditions that match the reported fragment velocities (Table 9) was roughly 9.7 MJ. This disagrees with energy calculated by this method. The disagreement may be caused by the assumption of the total mass being characterized by the average fragment velocity from which E was determined or by imprecision in orbital parameters.

Orbital-Debris Pattern for Ariane Breakup

An analysis was performed to determine the three-dimensional distribution of the debris in reference to the point of breakup for the Ariane tank. This distribution was determined by knowing the point in space that the third-stage tank broke up, forcing the orbits of the debris through that point, and then determining their distribution at a later time (about 10 s). The information used to determine the distribution was the latitude of the breakup, the distance of the third-stage tank from the center of the Earth at the time of the breakup, the mean motion, and the inclinations and eccentricities of the individual fragments. The effects of an osculating orbit were corrected, but the effects of drag were not.

Table 11. Predicted fragment velocity from the Delta NOAA and Delta Landsat caused by pressure bursts at near-design burst pressure (2750 kPa)

	Predicted velocity m/s	
Method	0.001-m-thick tank wall	0.002-m-thick tank wall
Pisces (Clam)	20–675	15–510
Baker et al.	611	538
Gurney	495	245

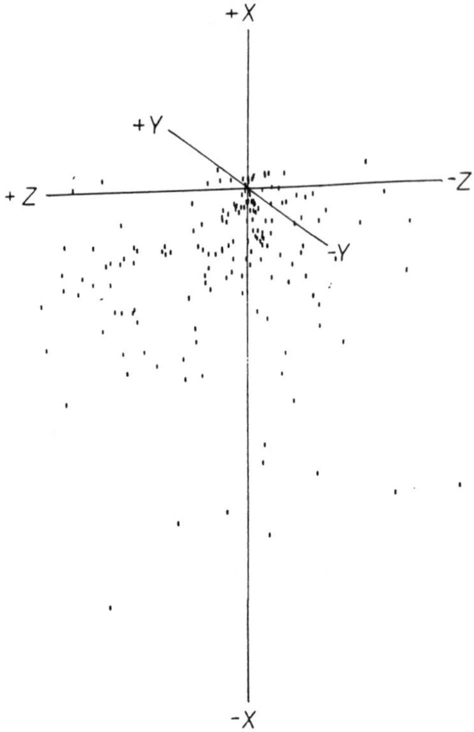

Fig. 7 Three-dimensional illustration of the orbital-debris pattern for the Ariane third stage.

A coordinate system for the third-stage tank was assumed to have a point at the center of the Earth; the "vernal equinox" was assumed to be 7 deg below the third stage. The assumed latitude and the distance of the third stage from the center of the Earth was determined from the orbital parameters given by Johnson (1987).[27] Based on this information, the effect of an osculating orbit was corrected. The right ascension of the ascending node for the fragments was determined based on the position of the third-stage tank and the inclination of the individual fragments. The true anomaly was determined using the distance to the center of the Earth and the eccentricity and mean motion of the individual fragments. The argument of perigee was determined from the true anomaly and the angle formed by the line of nodes and the third stage. The initial position of the fragments was assumed to be the position of the third stage when it exploded.

The position of the fragments as a function of time can be determined using the measured inclination, the corrected mean motion, and the calculated right ascension of the ascending nodes and argument of perigee. The fragment positions 10 s later were determined relative to the position of the fragment that is still identified as the Ariane third stage. These data are illustrated as two-dimensional plots in Fig. 6 where the + z axis is tangential to the original orbit of the third-stage tank and direction of original motion of the third-stage tank is from − z to + z. The center of Earth is in the − x axis. A three-dimensional plot of these figures is presented in Fig.7 with the same coordinate system.

The spacial distance from the origin of the breakup is also indicative of the velocity of the debris where large distances equate to large relative velocities. The debris with the largest space distance from the origin or large relative velocities are reported in the + z, + y, − x space, the + z, − y, − x space, and − z, − y, − x space. The debris in the − z, + y, − x space are characterized by lower spacial distance and velocity but have the highest concentration of low-velocity debris. Debris with velocities of any significance are not reported in any of the + x space. These debris spacial and velocity patterns are consistent with the Clam model where the original breach in the tank was directed toward the center of the Earth or − x direction.

Conclusions

Pressurized tanks containing residual propellants can undergo both physical and chemical explosions. Explosions resulting from physical pressure or from deflagrations produce similar tank failures (Clam model) with large-velocity distributions of the fragments. Explosions resulting from detonations produce a more uniform failure (Octant model) and small-velocity distributions of the fragments. The energy imparted to the fragments is usually less than 50% of the total energy of the explosion; the remaining energy expands the gas or vapor.

The propellants used in the Ariane third-stage tank, LH_2 and LO_2, can both deflagrate and detonate as gaseous mixtures and deflagrate and possibly detonate when mixed as liquids. The propellant used in the Delta second-stage tanks, A-50 and N_2O_4 will burn on contact. Detonation of mixtures of A-50 and N_2O_4 is believed unlikely because these propellants will react before they can establish mixtures capable of supporting stable detonations.

Review and analysis of data for the Ariane third stage indicate that the breakup of its tanks was caused by a physical pressure burst. This conclusion was supported by the magnitude and distributions of the fragment velocities and debris pattern and by the unlikely series of events required for deflagration or detonation of the propellants to account for the observed data. The initial failure of the tanks could have been from external sources, such as collision of small particles or a linear-shaped destruct charge, or from fatigue of the tanks and/or failure of the bulkhead that separates the tanks.

Review and analysis of data for the Delta NOAA and Delta Landsat second stages indicate that the tanks experienced a pressure burst near design burst pressure. The cause for the breakup is believed to be internal overpressurization of the tanks caused by leaks in the bulkhead and reaction of the A-50 and N_2O_4. The failure patterns of the tanks resemble those expected from slow overpressurization.

Acknowledgments

The authors wish to thank Meera Mukunda for the many hours she contributed running the Pisces code and discussing the overall problem. In addition, the assistance of Chris Barnet, who helped evaluate the orbital mechanisms, and of Cilia Carrillo, who generated the plots, is gratefully acknowledged.

References

[1]Baker, W. E., Kulesz, J. J., Ricker, R. E., Bessey, R. L., Westine, P. S., Parr, V. B., and Oldham, G. A., *Workbook for Predicting Pressure Wave and Fragment Effects of Exploding Propellant Tanks and Gas Storage Vessels,* NASA, Rept. 134906, 1977.

[2]Grelecki, C., *Fundamentals of Fire and Explosion Hazards Evaluation,* American Institute of Chemical Engineers, New York, 1976.

[3]Smith, I. D., Fritchman, T. T., Radwanski, C. F., and Bishop, C. V., "WSTF Tube Burst Tests," presented at the JANNAF Safety and Environmental Protection Subcommittee Meeting, NASA Lewis Research Center, Cleveland, OH, May 5-7, 1987.

[4]Strehlow, R. A., *Combustion Fundamentals,* McGraw-Hill, New York, 1984.

[5]Kinney, G. F. and Graham, K. J., *Explosive Shocks in Air,* Springer-Verlag, Berlin, 1985.

[6]Bodurtha, F. T., *Industrial Explosion Prevention and Protection,* McGraw-Hill, New York, 1980.

[7]Lewis, B. and von Elbe, G., *Combustion, Flames, and Explosions of Gases*, Academic, New York, 1961.

[8]Bishop, C. V., Benz, F. J., and Ullian, L. J., "Mixing of Cryogenic Fluids and Predicted Detonation Properties for Multi-Phase Liquid Oxygen and Liquid Hydrogen," paper presented at the JANNAF Propulsion Meeting, New Orleans, Aug. 26–28, 1986.

[9]Willoughby, A. B., Wilton, C., and Mansfield, J., "Liquid Propellant Explosion Hazards," URS Research Co. Burlingame, CA, AFRPL-FR-68-92, URS 652-35, 1968.

[10]Cook, M. A. and Udy, L. L., "Detonation Pressure of Liquid Hydrogen/Liquid Oxygen," Intermountain Research and Engineering Co., Salt Lake City, UT, Contract No. NAS8-5058, 1962.

[11]Deppe, G. and Plaster, M., "Liquid Oxygen/Liquid Hydrogen Explosion Hazards Characterization," NASA White Sands Test Facility, Las Cruces, NM, TP-WSTF-503, 1987.

[12]Fish, W. R., "Storable Liquid Propellants: Nitrogen Tetroxide/ Aerozine-50," Aerojet-General Corp., Sacramento, CA, LRP198, 1962.

[13]Bell Aerosystems Co., "Titan II Storable Propellant Handbook," Rev. B, Bell Aerosystems Co., Buffalo, NY, Contract No. AFO4(694)-72, Bell Rept. No. 8182-933004, 1963.

[14]Copeland, J. P., Gift, R. D., and Valentine, R. S., "The Determination of the Autodecomposition Characteristics of Aerozine 50," Atlantic Research Corp., Alexandria, VA, Final Rept. 37-5484, 1970.

[15]Pedley, M. D., Bishop, C. V., Benz, F. J., Bennett, C. A., McClenagan, R. D., Fenton, D. L., Knystautas, R., Lee, J. H., Peraldi, O., Dupre, G., and Shepherd, J. E., "Hydrazine Vapor Detonations," presented at the ICDERS 11th International Colloquium on Dynamics of Explosions and Reactive Systems and Tenth International Symposium on Combustion Processes, Warsaw, Poland, August 3-7, 1987; also published in *Dynamics of Explosions,* A. Kuhl et al., (Eds.) Vol. 114, Progress in Astronautics and Aeronautics series, AIAA, Washington, DC, Dec. 1988, pp. 45–63.

[16]Gordon, S. and McBride, B. J., "Computer Program for Calculation of Complex Chemical Equilibrium Compositions, Rocket Performance, Incident and Reflected Shocks, and Chapman-Jouguet Detonations," NASA Lewis Research Center, Cleveland, OH, NASA-SP-273, 1976.

[17]Zabetakis, M. G., "Fire and Explosion Hazards at Temperature and Pressure Extremes," presented at the A.I.Ch.E. I.E. Joint Mtg., London, June 13-17, 1965.

[18]Lee, J. H., "Physics of Gas Explosion (PH714)," INTEC Education and Training Division 3522, Sandia Labs., Albuquerque, NM, 1984.

[19]Pohl, W., Hoek, M. J. V. D., Buis, J. P., Florie, C. J. L., Hancock, S. L., Groenenboom, P. H. L., Hesselink, D. J. E., Lenselink, H., and Schaffers, P. J. J., *The PISCES Software for DEFENSE,* PISCES International B. V., The Netherlands, 1984.

[20]Su, S. Y. and Kessler, D. J., "Contribution of Explosion and Future Collision Fragments to the Orbital Debris Environment," *Advanced Space Research,* Vol. 5, No. 2, 1985, pp. 25–34.

[21]Kling, R., "Postmortem of a Hypervelocity Impact," Teledyne Brown Engineering, Colorado Springs, CO, CS86-LKD-001, 1986.

[22]Personal Communication (1987) from William Nance, R&D Associates, Marina Del Rey, CA.

[23]Arianespace, *Ariane 2 and Ariane 3 User's Manual,* Issue No. 1, Arianespace, 1981.

[24]Unpublished report included in personal communication of Dec. 22, 1986, from Remi Hergott written to Joseph P. Loftus Jr., NASA Johnson Space Center.

[25]McDonnell Douglas Astronautics Co., *Delta Spacecraft Design Restraints,* McDonnell Douglas Astronautics Co., Huntington Beach, CA, 1980.

[26]Personal communications (1987) with Andy Gray, NASA Goddard Space Flight Center.

[27]Johnson, N. L., *A Preliminary Analysis of the Fragmentation of the Spot 1 Ariane Third Stage,* Teledyne Brown Engineering, Colorado Springs, CO, TBE Tech. Rept. CS87LKD-003, 1987.

Chapter 6

Characteristics of Satellite Breakups from Radar Cross Section and Plane-Change Angle

Gautam D. Badhwar* and Andrew E. Potter†
NASA Johnson Space Center, Houston, Texas
and
Phillip D. Anz-Meador‡ and Robert C. Reynolds§
Lockheed Engineering and Sciences Company, Houston, Texas

I. Introduction

On-orbit satellite fragmentation has contributed to almost half of the objects orbiting the Earth. There are at least three major causes of fragmentation: propulsion-related explosions, intentional explosions, such as anti-satellite tests (ASAT), and collision or hypervelocity impacts. The resulting debris distribution depends on the mechanism of the breakup. Current observational techniques do not allow objects of less than about 8-cm diameter to be detected, even though objects as small as a few millimeters pose a potentially significant spacecraft hazard. An understanding of these objects has to be derived from information about the cause of fragmentation and from the fluxes of objects that can be seen (i.e., those with diameter above about 8-cm).

Culp and McKnight developed a Satellite Fragmentation Event (SAFE) Test that is based on analysis of mass distribution, variations of periods and inclination, trackable mass, and asymmetry of breakup in Gabbard plots.[1] This test assigns a score to six factors: 1) mass distribution exponential, 2) mass distribution describable by a polynomial of small finite order, 3) relative magnitude of dispersion of larger pieces of debris in the Gabbard plot, 4) asymmetry of dispersion, 5) orderedness of dispersion, and 6) change in velocity imparted to the larger pieces. Analytic and graphical approaches carry equal weight in the test scores. The graphical approach is inherently subjective; and with finite errors in the observed data, a true exponential distribution can be fitted with a polynomial of only a few

*Space Scientist.
†Supervisory Space Scientist.
‡Scientist.
§Senior Scientist.

terms. Criteria 1 and 2 in the SAFE test may therefore be redundant. In addition, the Gabbard plots are influenced by the age of the debris due to atmospheric drag. Although subjective, the SAFE test had a reasonable success when applied to 12 breakups. This suggested that a quantitative and objective approach was needed to yield a more satisfactory result.

This paper presents a methodology for characterizing the mechanism for on-orbit satellite fragmentation that is objective and provides the relative probability of the satellite breakup event due to any of three major causes: low-intensity explosion, high-intensity explosion, and hypervelocity collision. This is done by extracting simple features from the observed data to develop "signatures" that characterizes each of these classes. The distribution in fragment size (radar cross section) and orbital plane-change angle has been used to characterize the breakup events. A linear classifier is trained on a known set of breakup events to develop these signatures. The signatures of the unknown breakup satellites are then objectively compared to these signatures to classify the unknown events.

The approach described in this section is based on finding an analytic representation of both the radar cross section (RCS) and the orbital plane-change angle distributions of the breakup event. RCS estimates are made from the AN/FPS-85 radar operating at 442 MHz at Eglin Air Force Base, FL and the orbital element sets needed for the calculation of the plane-change angle are provided by radars operated by the U.S. Space Command (USSPACECOM). These radars have a threshold of about 8-cm diameter for objects orbiting at an altitude of 400 km. Thus only relatively large objects from breakup events can be detected. The parameters describing the frequency distribution of RCS and plane-change angle are estimated by fitting the observed data to a derived analytic form. A statistical analysis is used to select a subset of these features that contains most of the information about the type of breakup. These parameters are then used in a supervised classification mode to classify unknown breakup events.

In what follows, Section I describes the development of the models used to extract the relevant features from the RCS and plane-change-angle data. The RCS model is based on the experimental work of Bess.[2] Analysis of these data is discussed in Section III. In Section IV the features derived from this analysis are used in a supervised classification mode to classify various fragmentation events.

II. Model Development

Radar Cross Section (RCS)

Bess experimentally examined the mass distribution resulting from a hypervelocity impact, a high-intensity explosion, and a low-intensity explosion and showed that the cumulative mass distribution $N(>m)$ of the debris caused by a hypervelocity (3.0–4.5 km/sec) projectile impact takes the form:

$$N(>m) = N_0 m^{-\nu} \tag{1}$$

where N_0 and ν are two constants, whereas the distribution function for explosions followed an exponential distribution in the square root of mass:

$$N(>m) = N_0 \text{ Exp } (-c\sqrt{m}) \tag{2}$$

where c is a constant.

Equation (2) can be converted to a differential frequency distribution in radar cross section R by first relating the mass m to area of the fragment A, as $m = b\,A^{\delta}$ where δ, for example, varies between 0.5 for a uniform plate to 0.75 for a uniform sphere, and b is related to the density. This gives:

$$\frac{dN}{dA} = \frac{N_0 c\sqrt{b}}{2} A^{\left(\frac{\delta}{2}-1\right)} \text{ Exp } (-c\sqrt{b\,A^{\delta}}) \tag{2a}$$

For the Mie scattering region, R/A is a function of the ratio $2\pi r/\lambda$, where λ is the operating wavelength of the radar and r is the radius of scattering sphere. This function is quite complicated; but for $r<0.1\lambda$ (Rayleigh region) it has a slope of 3 in A and for $0.1\lambda<r<1.6\lambda$ (Mie region) an average slope of about 1. Combining Eq. (2a) with the expression for Mie scattering suggests the differential distribution of the radar cross section R can be written as follows:

$$\frac{dN}{dR} = K\,R^{\alpha} \text{ Exp } (-\beta R^{\gamma}) \tag{3}$$

where α, β, and γ are constants. Note that for $\beta = 0$ this equation is a power law; for $\alpha = 0$, an exponential law. Normalizing, the equation such that

$$\int_0^{\infty} \frac{dN}{dR} = 1$$

(note that peak RCS value is reached at the point R_m) yield:

$$\beta = \left(\frac{\alpha}{\gamma}\right) \frac{1}{R_m^{\gamma}} \tag{4}$$

Then the distribution function can be written as follows:

$$\frac{dN}{dR} = \frac{\gamma\left(\frac{\alpha}{\gamma}\right)^{\frac{\alpha+1}{\gamma}}}{\Gamma\left(\frac{\alpha+1}{\gamma}\right)} \frac{1}{R_m} \left(\frac{R}{R_m}\right)^{\alpha} \text{ Exp } \left\{-\left(\frac{\cdot\alpha}{\gamma}\right)\left(\frac{R}{R_m}\right)^{\gamma}\right\} \tag{5}$$

A derivation identical to the one followed above using Eq. (1) again leads to a model of the form of Eq. (3). Thus these three kinds of breakup events can be characterized in the RCS distribution by α, R_m and γ.

Physically these parameters represent the asymptotic rise, the peak, and the asymptotic fall in the differential RCS distribution. It should be remembered that implicitly these parameters depend on the physical density of the satellite fragments, the operating radar wavelength, the altitude, and the age of the debris fragments. A simple analysis shows that, drag being proportional to area, the differential distribution will shift to higher values of α, and γ and lower value of R_m as atmospheric drag removes debris from the orbit. A more refined analysis should take these factors into account and analyze the data as close to the time of breakup as possible. Equation (5) has been used in the analysis of the all of the RCS data.

Plane Change Angle Distribution

One of the important characteristics of a breakup process in orbit is the velocity perturbations applied to the fragments during the breakup. The velocity perturbation will, in general, change energy, angular momentum, and the plane of the orbit. Of all of the orbital elements that define the orbit of a given debris fragment, the inclination changes the least for the major perturbations encountered in low Earth orbit. If the inclination of the object before breakup is known, and the latitude at which it occurred is also known, the plane-change angle, which essentially measures the crossrange velocity perturbation, can be determined, and this can be done using either data taken immediately after the breakup or "old" data; the quantities needed to define plane change angle do not change with time.

The spherical triangles ABC and A′BC in Fig. 1 show the relationship between the plane-change angle θ, the initial and final inclinations i_I and i_F, the latitude of the breakup 1, and the arguments of latitude θ_I, and θ_F for a breakup occuring at B, with fragments being injected from an initial to a final orbit at this point:

$$C_\theta = \frac{C_{i_I}C_{i_F} + S_{i_I}S_{i_F}C_{\theta_I}C_{\theta_F}}{1 - S_{i_I}S_{i_F}S_{\theta_I}S_{\theta_F}} = \frac{C_{i_I}C_{i_F} + \sqrt{S_{\theta_I}^2 - S_1^2}\,\sqrt{S_{i_F}^2 - S_1^2}}{1 - S_1^2} \qquad (6)$$

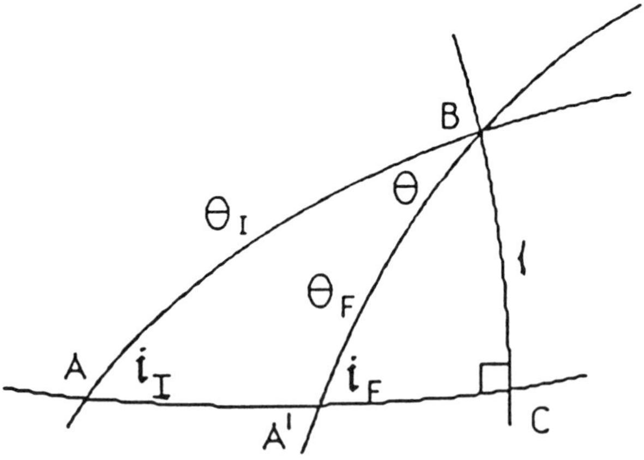

Fig. 1 The spherical triangles showing the geometry of the plane-change-angle calculation. C and S: sine and cosine of angle, respectively.

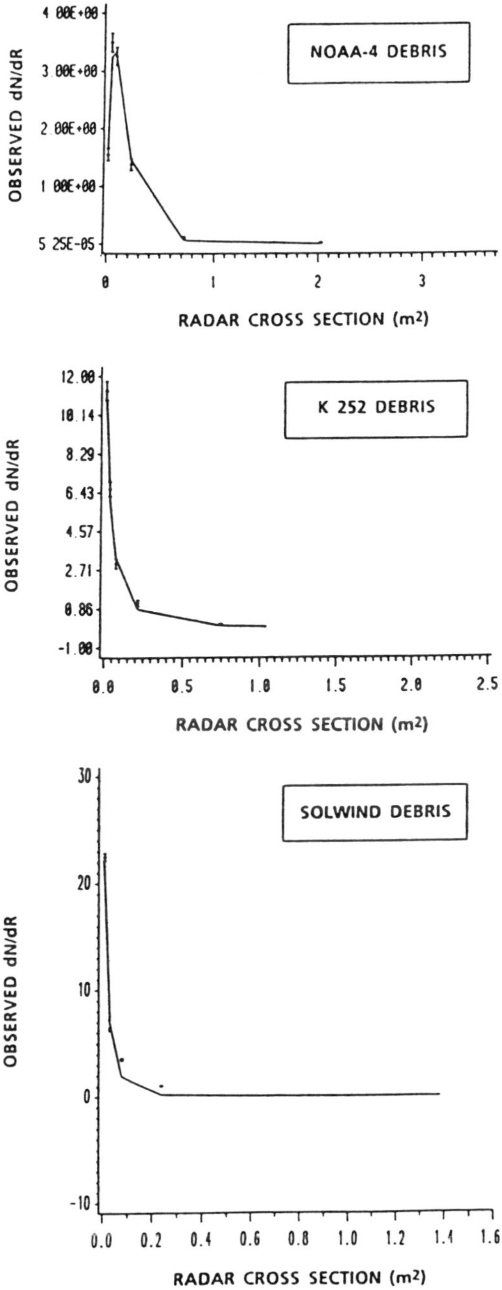

Fig. 2 Frequency distribution of weighted RCS for breakup of NOAA-4 (low-intensity explosion), Cosmos 252 (high-intensity explosion), and Solwind (hypervelocity impact). Points with error bar, actual observed values; solid curve, best fit to Eq. (5).

where $S_{\theta_I} = S_1/S_{i_I}$ and $S_{\theta_F} = S_1/S_{i_F}$. Although θ_I and θ_F cannot be determined unambiguously knowing only the latitude of the breakup, both $S_{\theta_I}S_{\theta_F}$ and $C_{\theta_I}C_{\theta_F}$ will be greater than zero in all cases except where the argument of latitude is very near 90 or 270 deg or where the velocity perturbations are extremely large.

For time-invariant calculation of the plane-change angle, the latitude of the breakup event must be known; this information is taken from the work of Johnson et al.[3]

The insensitivity to time in the calculation of plane-change angle is important because data on fragmentation events are often not available until long after the breakup occurs. However, beyond the fact that the quantities in Eq. (6) do not vary in time, the plane-change angle, because it is measuring the cross-range velocity perturbation and has only a second-order effect on the energy, is less susceptible to selection effects than other characteristics, such as change in semi-major axis.

Wiesel first suggested that the plane-change angle can be used to separate the type of explosion.[4] Based on simple arguments of the equipartition of energy and a spherical explosion geometry, he showed that the distribution should be a gaussian. An examination of the differential-plane-change-angle distribution $dN/d\theta$ as a function of the plane-change angle θ for a number of breakups showed that the distribution is essentially a gaussian but with constant background contribution. This distribution can be parametrized as follows:

$$\frac{dN}{d\theta} = \frac{A}{\sqrt{2\pi\sigma^2}} \operatorname{Exp}\left[-\frac{(\theta - \theta_p)^2}{2\sigma^2} \right] + \eta \tag{7}$$

where θ_p is the peak of the distribution, σ is its variance, and η is the background level.

III. Analysis

Radar Cross Section

RCS data were grouped into ten nonuniform intervals, and the weighted RCS values of each interval and the number of fragments falling in each interval were calculated. The grouped data were fitted to the model of Eq. (5) by a nonlinear fitting algorithm to estimate the three parameters α, R_m, γ and their associated errors.

Figure 2a shows the observed data for the NOAA-4 breakup, the consequence of a propulsion-related low-intensity explosion. The points with error bars are the data points, and the solid line is the best-fit curve that joins the predicted value estimated by the best fit to Eq. (5) of the observed distribution. Figure 2b shows a similar plot for an alleged Soviet ASAT test, Cosmos 252 , destroyed by a high-intensity explosion.[3] The Cosmos 252 curve is quite steep compared to the NOAA-4 breakup. Figure 2c plots the only definitely known hypervelocity-impact breakup (Solwind); the frequency distribution resembles the Cosmos 252 breakup more than the NOAA-4 breakup. Figure 3 plots the 226 fragments from the breakup (cause unknown, although explosion most likely) of the Ariane second

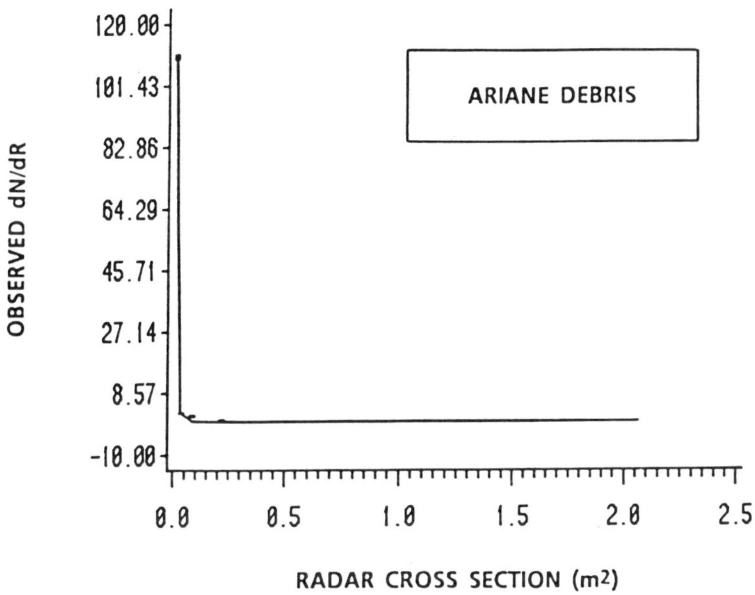

Fig. 3 Frequency distribution of the weighted RCS for breakup of Ariane second stage. Note the large number of pieces at the threshold of observation of the radar. The cause of this breakup is under investigation; it is believed to be a high-intensity explosion.

stage. It shows that the distribution may be even steeper than the Cosmos 252 breakup. As was expected from the earlier discussion , the RCS distribution can indeed be described by the model of Eq. (5).

Figure 4 plots α, R_m, and γ for all of of the breakups for which RCS data are available. The symbols identify the various parent satellites; the shades identify the proposed causes of breakups. Much of this information is derived from Johnson et al.[3]

All the low-intensity explosions lie substantially above the α, γ plane. It should also be pointed that the values of γ and α observed are roughly in qualitative agreement with values for explosions as seen in the work of Bess.[2] They are also in accord with the Bess's observation that c for a high-intensity explosion is greater than for a low-intensity explosion . Figure 4 suggests a rough classification scheme: collision-type breakups are defined by $R_m \sim 0$, high-intensity explosions by $R_m \sim 0$, $\alpha > \gamma$, and low-intensity explosions by $R_m > 0.05$ and α, $\gamma > 0.3$.

The technique just described differs substantially from the work of Culp and McKnight,[1] in that they work with the cumulative mass density distribution and thus require a knowledge of the spacecraft density or, more precisely, the fragment densities and that they do not consider the effects of Mie scattering on the relation between size and RCS. They also do not have a single model to describe all types of breakup events.

In short, it has been shown that the RCS data can be fitted to a single three-parameter model that has experimental support and that these parameters permit a rough separation of breakup types.

Plane-Change-Angle Analysis

The plane-angle data for each breakup were grouped into 0.18-deg intervals, and the differential number density $dN/d\theta$ was computed to unity of the total area under the curve. This normalized distribution was fitted to the model of Eq. (7) and the parameters corresponding to the peak of the distribution, its variance, and the background values estimated. The accuracy with which the plane-change angle can be computed depends on the precise knowledge of the breakup latitude and longitude, which if accurately known gives accuracy in plane change to better than 0.1 deg. As already mentioned, the plane-change angle is relatively insensitive to age of the breakup.

Figure 5a shows the plane-change-angle, distribution for the NOAA-4 fragmentation. The error bars again correspond to the number of fragments

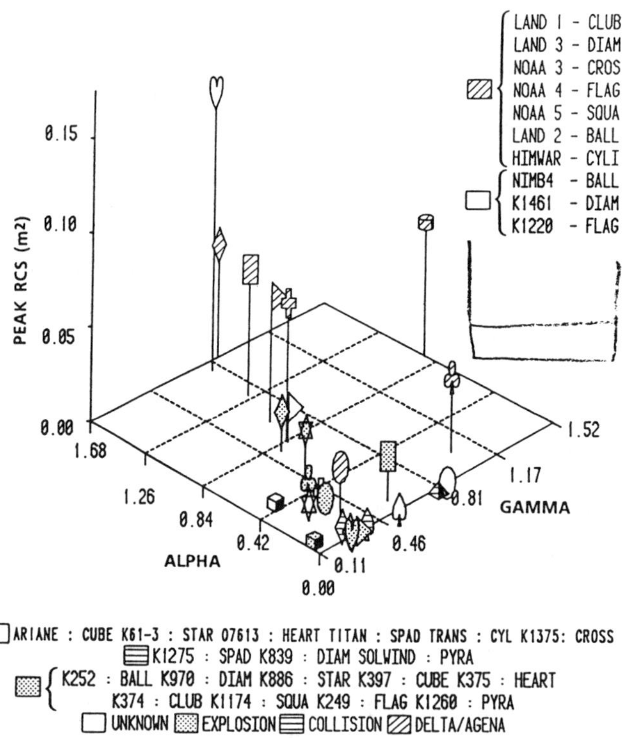

Fig. 4 Three-dimensional plot of parameters derived from RCS data for the 26 breakup events. High-intensity explosion shown as dots; hypervelocity impacts, as horizontal stripes; low-intensity explosion, as cross stripes; and unknown events, as blank. Note low-intensity explosions lie above the α, γ plane; general clustering of the high-intensity explosion data.

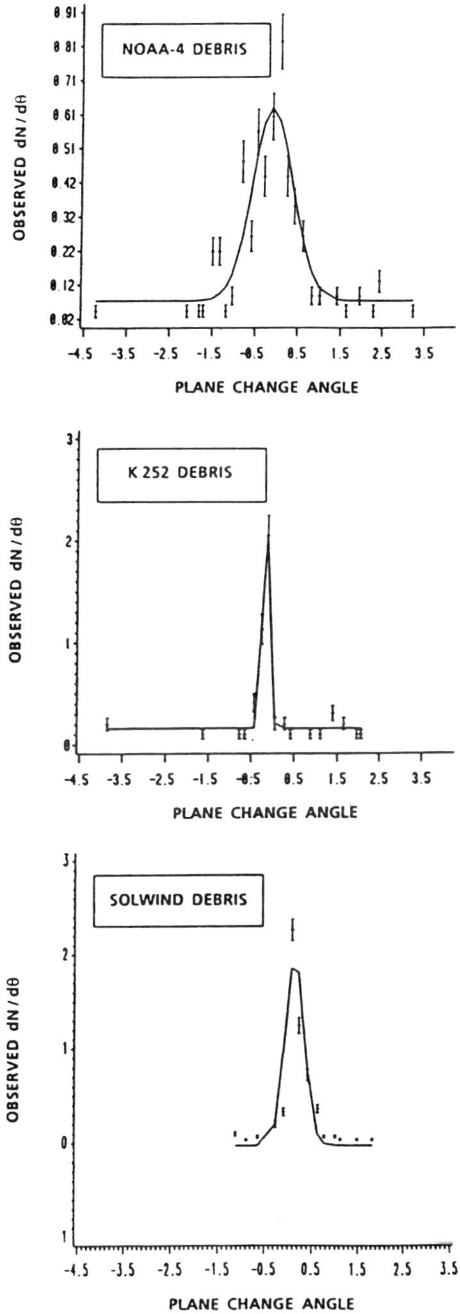

Fig. 5 Frequency distribution of the plane-change angle for breakup of NOAA-4 (low-intensity explosion) Cosmos 252 (high-intensity explosion), and Solwind (hypervelocity impact). Points with error bars, observed values; solid curve, best fit to gaussian Eq. (7).

in each interval. Fluctuations reflect a very fine interval size. The solid curve is the best fit to the data using Eq. 6. Figures 5b and 5c show similar plots for Cosmos 252, an alleged ASAT high-intensity explosion, and Solwind, the only definitely known hypervelocity-impact breakup. The Cosmos 252 and Solwind breakups have narrow distribution compared to NOAA-4's. This is what one might expect because a low-intensity explosion would, like an overpressurized balloon, more likely scatter the debris into a large angular range than would a high-intensity explosion. Figure 6 shows the same plot for the debris from the Ariane V16 breakup. Because of the smaller range of data, the interval is 0.05 deg. This distribution is extremely narrow and much more like a high-intensity explosion than a low-intensity one. The secondary peaks on either side of the central peak are asymmetric. The variance of these peaks is quite comparable to that of the central peak. An examination of the RCS data of these secondary peak shows no particular difference from the RCS distribution of the main peak. The reasons for these peaks are not understood but may well be the accuracy with which θ can be calculated. In subsequent analysis here, the variance of the central peak distribution will be used.

Figure 7 plots the variance, the peak, and the background level of the plane-change-angle distribution for all of the breakup events. The high-intensity explosions are all clustered around low-variance values, and low-

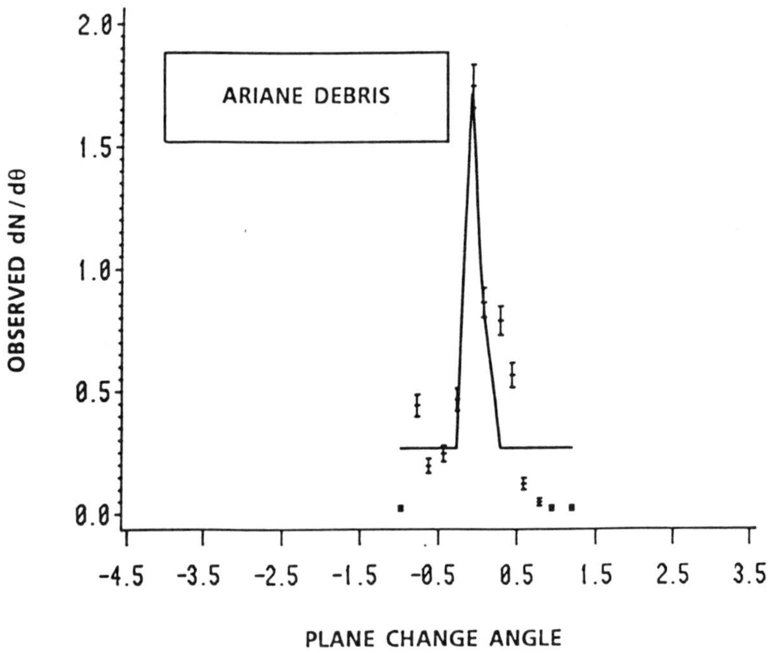

Fig. 6 Frequency distribution of plane-change angle for breakup of Ariane second stage. The fit to data is not as good as for NOAA-4 and Cosmos 252 breakups. However, the central peak is very narrow, suggesting a high-intensity explosion.

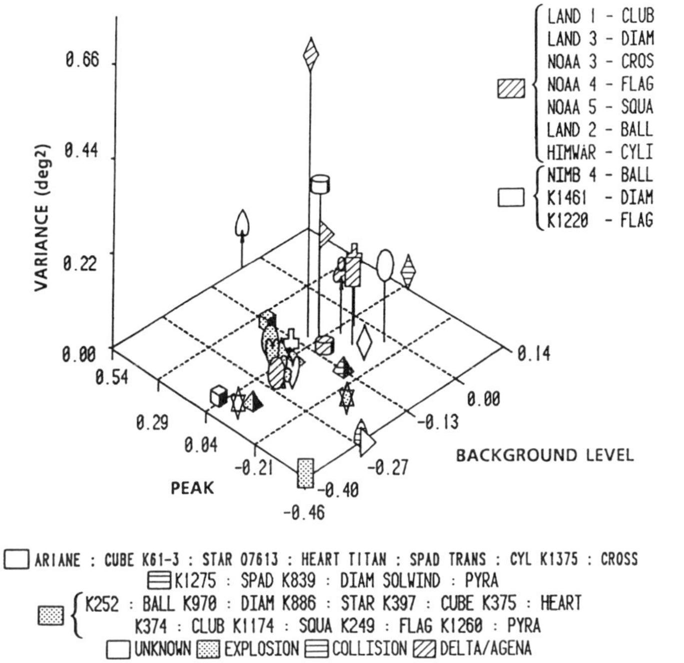

Fig. 7 Three-dimensional plot of parameters derived from fitting the plane-change-angle frequency distribution for the breakup events . Note that location of the peak of the plane-change-angle distribution provides no separation into classes and much of the separation information is contained in the measure of the variance.

intensity explosions around high-variance values. The exception, the Landsat 2 upper-stage breakup, was characterized by two breakup events and should have been treated as such. As expected, the peak values do not allow any separation because it tells more about which way the momentum transfer took place and not why the breakup occurred. The background values do provide some separation; but the error in its determination is typically large because of the poor statistics at the tail ends of the distribution.

Analysis of the Combined RCS Plane-Change-Angle Data

A principal component analysis of the all of the derived features shows that the four features α, γ, R_m, and σ carry most of the information on the class of the breakup event. It should be emphasized that other features derived from the angular-momentum-transfer, energy, or velocity perturbations may be more powerful discriminators of breakup type but have not been carefully analyzed.

From Figs 4 and 7 it can be seen that neither the RCS-derived parameters nor the plane-change-angle-derived parameters individually suffice to separate the type of breakup. Furthermore, only the variance of the plane-

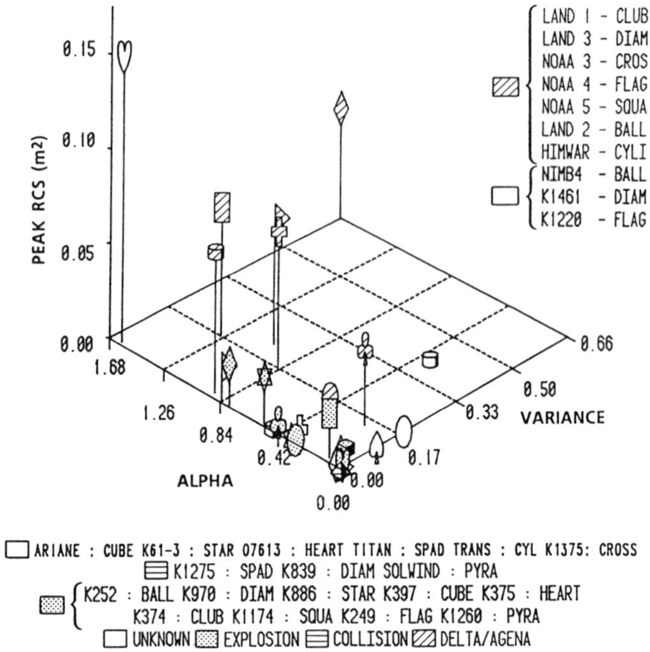

Fig. 8 Three-dimensional plot of parameters σ, α, γ for breakups. From Figs. 4 and 7 it is clear that information from RCS or the plane-change angle by itself is not sufficient for class separation, but combining the two gives a better measure of separation.

change angle is important among the plane-change-derived parameters. However, combining the two sets of parameters does lead to a better separation, as shown in Fig. 8, a plot of the σ, α, and R_m. Figure 8 shows that the separation into three classes is indeed better than that indicated in either Fig. 4 or 7.

IV. Classification of Breakup Events

For the set of breakup events containing the variables α, R_m, γ, and σ and a classification variable defining the group of breakups (high-intensity explosion, low-intensity explosion, and hypervelocity impact), a linear discriminant function or classification criterion can be defined by a measure of the generalized squared distance[5]:

$$D_t^2(\boldsymbol{x}) = g_1(\boldsymbol{x},t) + g(t) \tag{8}$$

where $\boldsymbol{x}$ is the vector of observations and t a subscript to distinguish the group and—

$$g_1(\boldsymbol{x},t) = \overline{(\boldsymbol{x} - \boldsymbol{\mu})}\ \Sigma_t^{-1}\ (\boldsymbol{x} - \boldsymbol{\mu}) + 1n\,|\Sigma_t| \tag{9}$$

and $g(t) = 0$ if all classes are equally probable and $= -1l$ (the prior probability for group t), μ is the vector containing the means of the variable in the group t, and Σ is the covariance matrix.

The posterior probability of an observation x belonging to the group t can be expressed as follows:

$$p_t(x) = \frac{\text{Exp}\ [-0.5\ D_t^2(x)]}{\Sigma_t\ \text{Exp}\ [-0.5\ D_t^2(x)]} \tag{10}$$

An observation belongs to the group $t = k$ if $D_k^2(x)$ if the least or $p(x)$ is the highest value.

The discrimination procedure was trained on breakup events of known category, and the same observations reclassified. This gives the best possible performance of the classifier. It should be noted that the accuracy of any such classifier depends on how well the "training class" samples the actual class distribution. Since there are only three known or suspected hypervelocity-impact events, this increases the uncertainty with which this class can be identified.

The division of all the breakup events into these three classes is based on the work of Bess[2]; but it is quite arbitrary in the sense that each explosion is unique in its energetics and where the detonation charge is located on the body. Similarly , the collision can be head-on or glancing. Thus in each class there exists a distribution of parameters that describes that overall class. This is only a first-order, broad classification.

Table 1 gives the confusion matrix, and Tables 2 and 3 give the actual break-down of the posterior membership probabilities. Cosmos 375 is misclassified. It is believed that the orbital elements of the fragments from the breakup of Cosmos 374 and Cosmos 375 have been mixed (Johnson, private communication, May 1987). Using the training statistics developed, the breakup events labeled black (hollow) were classified.

A comparision of these results with the work of Culp and McKnight shows good agreement in two class separation—low-intensity explosion (called propulsion-related by Culp and McKnight) and (high-intensity + hypervelocity) breakups. Nimbus 4. although thought to be a propulsion-related breakup, is not classified as propulsion-related in this study or in the Culp and McKnight's analysis. However, Cosmos 1275 is considered to be a hypervelocity impact by Culp and McKnight but this analysis shows only a 38.1% probability for impact. OPS 7613 is a Delta-class breakup, and is so classified. In this analysis, Cosmos 1461 is misclassified as a collision; it should be considered a high-intensity explosion.[3] In this case, there were two distinct breakup events that occurred at different times. They have been treated as one in this analysis. A more refined analysis would treat them as separate events; hence the misclassification.

The analysis of the Ariane V16 breakup shows that it was a high-intensity explosion with a probability of 89.6% and a collision with a probability of 10.4%. If one believes that Cosmos 1275 is a collisional breakup and the data reclassified, Ariane would show probabilities of 78.6% and 21.4% respectively for same two categories. It was mentioned earlier that there are

Table 1 Classification of debris by known data—Number of observations and percents classified into type

Type	C	D	E	Total
C	3	0	0	3
	100.00	0.00	0.00	100.00
D	0	7	0	7
	0.00	100.00	0.00	100.00
E	1	0	6	7
	14.29	0.00	86.71	100.00
Total	4	7	6	17
Percent	23.53	41.18	35.29	100.00

Note: C = hypervelocity impact; D = low-intensity explosion; E = high-intensity explosion.

Table 2 Classification of debris known data—Posterior probability of membership type (from type in Table 1)

Satellite	From type	Classified into type	C	D	E
Cosmos 839	C	C	0.6031	0.0000	0.3969
Solwind	C	C	0.9185	0.0000	0.0815
Cosmos 1375	C	C	0.6988	0.0000	0.3012
Hiwari	D	D	0.0000	1.0000	0.0000
Landsat 1	D	D	0.0000	1.0000	0.0000
Landsat 3	D	D	0.0000	1.0000	0.0000
NOAA 3	D	D	0.0000	1.0000	0.0000
NOAA 4	D	D	0.0000	1.0000	0.0000
NOAA 5	D	D	0.0000	1.0000	0.0000
Landsat 2	D	D	0.0605	0.9034	0.0361
Cosmos 252	E	E	0.3228	0.0000	0.9772
Cosmos 970	E	E	0.0826	0.0000	0.9174
Cosmos 886	E	E	0.2141	0.0000	0.7858
Cosmos 397	E	E	0.4724	0.0000	0.5276
Cosmos 375	E	C[a]	0.6205	0.0000	0.3795
Cosmos 374	E	E	0.2166	0.0000	0.7834
Cosmos 1174	E	E	0.0377	0.0000	0.9623

[a]Misclassified observation (probably due to mixing of K 374 and K 375 orbital elements).
Note: C = hypervelocity impact; D = low-intensity explosion; E = high-intensity explosion.

secondary peaks on either side of the central peak in the plane-change-angle differential distribution and that the variance of the central peak was used in the analysis. The results above are not affected if the data is sorted into coarser bins and refitted to smooth out the secondary peaks, because this distribution is extremely narrow to begin with.

This conclusion differs from an analysis of Benz et al.[6] (see Chapter 5). They conclude that the breakup of the Ariane third stage was caused by a physical pressure burst; that is, a pressure not from a chemical reac-

Table 3 Classification of debris unknown data—posterior probability of membership type

Satellite	From type	Classified into type	C	D	E
Ariane	U	E	0.2142	0.0000	0.7858
Cosmos 1275	U	E	0.3813	0.0000	0.6187
Cosmos 61-63	U	F	0.3147	0.0000	0.6853
Nimbus 4	U	C	0.9328	0.0000	0.0672
OPS7613	U	D	0.0000	1.0000	0.0000
Titan 3C	U	C	0.8198	0.0000	0.1802
Transit	U	C	0.8816	0.0005	0.1178
Cosmos 1461	U	C	0.5510	0.0000	0.4490
Cosmos 1220	U	E	0.1358	0.0000	0.8642

tion—e.g., inert gas pressure. This analysis was based on a comparison of the predicted fragment velocities based on models of exploding orbital tanks to the fragment velocities calculated from the observed orbital elements of the fragments, an analysis of the debris pattern, and a comparison of the total energy calculated for the debris with total energy calulated for the explosion based on theoretical models. Their calculation assumed that the radar observed most of the debris and the orbital parameters are known with great precision. Both of these assumption are not quite correct. Indeed, the radar does not see about two-thirds of the fragments. Benz et al. do suggest that the initial failure of the tanks could have been due to a collision with small particles, the material from a linear-shaped destruct charge, or from fatigue of the tanks and/or failure of the bulkhead that separates the fuel tanks. Our results can be consistent with the analysis of Benz et al. if the physical-pressure-burst mechanism advocated by them produces a extremely narrow plane-change-angle distribution. If so, the definition of high intensity explosions suggested by Bess needs to be enlarged to encompass such events.

Aerospaciale stated that Ariane engineers had concluded that a possible cause of the breakup was leakage from the hydrogen tank.[6] When pressure in the hydrogen tank dropped below that in the oxygen tank, the bulkhead separating the two chambers reversed and broke, causing a catastrophic breakup of the entire structure.

V. Conclusions and Recommendations

A three-parameter model has been developed that describes the differential RCS distribution. It has been related to laboratory-derived mass distributions for various breakup categories, and thus permits a classification for various satellite breakups. However, these parameters depend on "age" and fragment density.

Moreover, a parametric representation has been found for the plane-change-angle differential distribution. The variance of this distribution is an effective separator of breakup classes.

A linear discriminant procedure trained on known breakups has been used to classify unknown breakups. Although they do differ in specific cases, the classifications are in general agreement with the work of Culp and McKnight.

Additional information derived from the orbital elements such as angular momentum, energy change, and decay rates should be examined to refine the classification results. The "age" dependance of the RCS data distribution should also be taken into account.

Furthermore, we see a clear need to identify additional hypervelocity-impact events. The signatures of such events have been very firmly established.

References

[1]Culp, R. D. and McKnight, D. S., "Analysis of the Origins of Debris Clouds," Teledyne Brown Engineering, CR SC 7460, Oct. 1986.

[2]Bess, T. D., "Mass Distribution of Orbiting Man-Made Space Debris," NASA Langley Research Center, Hampton, VA, Rept. L-10477, Dec. 1975.

[3]Johnson, N. L., Gabbard, J. R., Kling, R. L., Jr., and Jones, T. W., "History of On-orbit Satellite Fragmentation," Teledyne Brown Engineering, Tech. Rept. CS 86-USASDC-001, Feb. 1986.

[4]Wiesel, W., "Fragmentation of Asteriods and Artifical Satellites in Orbit," ICARUS, Vol. 34, 1978, pp. 99–116.

[5]Rao, C. R., *Linear Statistical Inference and its Applications,* Wiley, New York, 1973.

[6]Private communication to Joseph Loftus, NASA Johnson Space Center, Houston, TX, July 1987.

Chapter 7

Albedo Estimates for Debris

A. E. Potter* and Karl G. Henize†
NASA Johnson Space Center, Houston, Texas
and
D. L. Talent‡
Lockheed Engineering and Sciences Company, Houston, Texas

Introduction

Almost half of the objects tracked by the U. S. Space Command (USSPACECOM) are fragments from the breakup of larger satellites. Satellite breakups produce many more small fragments than large ones. Since objects smaller than 10 cm in diam are not tracked by USSPACECOM, the actual debris environment includes a great many more particles than are listed in the satellite catalog published by USSPACECOM. And since particles as small as 1 mm in diam can pose a significant collision hazard, it is necessary to estimate the population of the small debris from these breakups.

The size distribution of debris from a breakup depends on its cause. Many rocket upper stages have broken up as a result of the explosion of residual fuel and oxidizer. Deliberate detonation of an explosive charge has been used to destroy military satellites after their useful lifetime. Collision with another satellite at hypervelocity can also cause a breakup. How a breakup occurred determines distribution of particle size. Thus, to model the debris environment requires knowledge of the cause of each breakup.

Determining the albedo of fragments might do so. Laboratory studies[1] of the impact of hypervelocity projectiles on simulated spacecraft has shown that the fragments from an impact have a thin soot layer, apparently from breakdown of plastics in circuit boards on the simulated spacecraft. Fragments produced from detonation of an explosive charge might also be blackened by soot from the explosive, while fragments resulting from a pressure vessel explosion would be expected to be soot-free. Clues as to the

*Chief, Space Science Branch, Solar System Exploration Division.
†Senior Scientist, Space Science Branch, Solar System Exploration Division.
‡Scientist, Orbital Debris Group.

cause of a satellite breakup might thus be found from measurements of particle albedo.

Another (perhaps more important) reason for interest in debris albedo has to do with the interpretation of optical measurements of small debris. Telescopic measurements have been performed to define the debris population at sizes below the reach of USSPACECOM radars. The 31-in. telescopes operated by MIT Lincoln Laboratory at Socorro, New Mexico, can detect satellites much smaller than the limiting, 10-cm size of USSPACECOM. If the particles have an albedo of 0.5, then the Socorro telescopes can detect particles as small as 1 cm at an altitude of 500 km. However, the physical size of particles determined from optical brightness depends on the value assumed for the albedo. Smaller albedos would lead to a larger limiting size for telescopic detection.

Consequently, to interpret the optical data, it is essential to have accurate information about the albedo of debris particles and the various processes that may affect albedo.

This work thus measured the albedo of a number of debris fragments from various satellite breakups, seeking to determine whether such information provides a unique identification of the cause of a breakup.

Approach

Optical magnitudes were measured for orbiting fragments from various breakups. Debris from the breakup of the third stage from Ariane V19, which launched the SPOT-1 and Viking payloads, was the primary object of investigation, since this breakup was an especially severe one, occurring at a high altitude, which guaranteed that the debris would remain in orbit for decades. Because this satellite was in a Sun-synchronous orbit, debris could be observed only at high northern latitudes in summer, and vice versa in the southern summer. Debris from Cosmos 1275, two Delta third stages, and Solwind (the P78 satellite) were measured for comparison.

The cause of the Ariane V21 breakup was unknown. The Cosmos 1275 breakup was suspected to be the result of a collision, and the Delta third-stage breakups were believed to be caused by explosion of residual fuel and oxidizer. A collision between an antisatellite weapon and the satellite caused the Solwind breakup.

Satellites Observations

Observations were made between January 6 and 15, 1987, at the Black Birch Observatory of the U.S. Naval Observatory, located near the town of Blenheim on the South Island of New Zealand. The instrument was a 20-cm-aperture, f/1.3 catadioptric telescope equipped with an intensified vidicon sensor. This system (1.9- by 2.1-deg field of view) can detect a 10th-magnitude satellite moving at 0.4 deg per second.

For each night's observing, look angles were calculated using the Teledyn-Brown SATRAK program, which utilizes USSPACECOM orbit elements. The telescope was set on the predicted coordinates of the satellite culmination point 2–3 min before the passage time. Field centering was verified using star fields generated by a portable Compaq computer utilizing

Table 1 Satellite observations in the Ariane campaign

Original satellite	Date of breakup	Probable cause	Number of satellites Detected	not seen	Total
Ariane 3rd Stage	Nov. 13, 1986	Unknown	15	4	19
Landsat-1	May 22, 1975	Explosion	8	2	10
Landsat-3	Jan. 27, 1981	Explosion	4	3	7
Cosmos 1275	July 24, 1981	Unknown (collision?)	17	5	22
Solwind (P-78)	Sept. 13, 1985	Collision	8	11	19

the Smithsonian Astrophysical Observatory (SAO) star catalog.

Table 1 summarizes the number of objects observed from each breakup. Although objects with radar cross section (RCS) smaller than 0.8 are seldom observable, objects as small as RCS 0.3 have been included in the observing list. For the "not seen" objects, it was verified in the course of data reduction that these were not missed due to poor orbital elements or weather. In these cases, an assumed limiting magnitude has been used to obtain an upper limit to the albedo.

Data Analysis

A Gould image-analysis system was used for data intepretation. Calibration of the image data in terms of stellar magnitude was done in several steps. The first step was to establish a calibration based on measurements of standard stars in the selected areas—regions containing well-established photometric sequences. For this, a video frame of the selected area was digitized. Linear profiles were plotted through star images and through the sky background immediately adjacent to them. The sky profile was subtracted from the star-plus-sky profile to obtain an intensity profile through the star alone. This profile was integrated to obtain a total intensity for the star.

A number of stars were measured to establish a curve of total intensity versus stellar magnitude. In effect, video gain and atmospheric haze translated this calibration curve up or down, so that the correct calibration curve could be established for a video frame with a different gain or extinction condition by measuring calibration stars in the frame and determining a correction factor from that measurement compared to the original calibration curve. Another factor which had to be taken into account for satellites was angular motion of the satellite across the field of view. This was calibrated by measurement of changes in the total intensity of standard stars as they were moved across the field of view by slewing the video telescope at different rates.

To analyze the satellite-image data, a video frame which included the satellite was digitized, and the satellite intensity was measured by plotting a linear profile through the satellite image and subtracting the background intensity measured adjacent to the satellite image. Three stars from the SAO catalog of known magnitudes were also measured in the same digitized frame. These values, averaged, were used to determine the offset from the original intensity-magnitude calibration curve. The angular velocity of the satellite was determined by measurement of the angular distance moved by the satellite through several frames. This value was used to correct the calibration curve to account for angular motion. Finally, the magnitude of the satellite was calculated, using the calibration curve corrected for video gain, extinction, and satellite motion. In many cases, the intensity of the satellite varied as it moved across the field of view, presumably due to tumbling. If a variation was noted, maximum, minimum, and mean values of the satellite intensity were measured, along with the time between maxima or minima. For the purposes of this analysis, only the mean intensity values were used.

To compare the satellite magnitudes, they were reduced to a standard phase angle and distance, chosen to be 90 deg and 1000 km, respectively. For each observation, the phase angle (angle at the satellite between the Sun and the observer) and the distance were provided by the original SATRAK look-angle calculations. (SATRAK is a computer program which generates look angles and related information from USSPACECOM orbital-element data.) Magnitudes were corrected to a solar phase angle of 90 deg (half solar illumination) using a Lambertian phase function, and to a range of 1000 km assuming an inverse square law.

Calculation of Satellite Albedos

To calculate the satellite albedo from the optical magnitude, it is necessary to know the cross-sectional area of the satellite. The only data available for the satellites measured in this work are the radar cross sections published by the USSPACECOM in the Satellite Situation Report. The RCS does not necessarily correspond to the geometric cross section. It was therefore considered desirable to perform a brief investigation of the relation between RCS as measured by USSPACECOM and the geometric cross section.

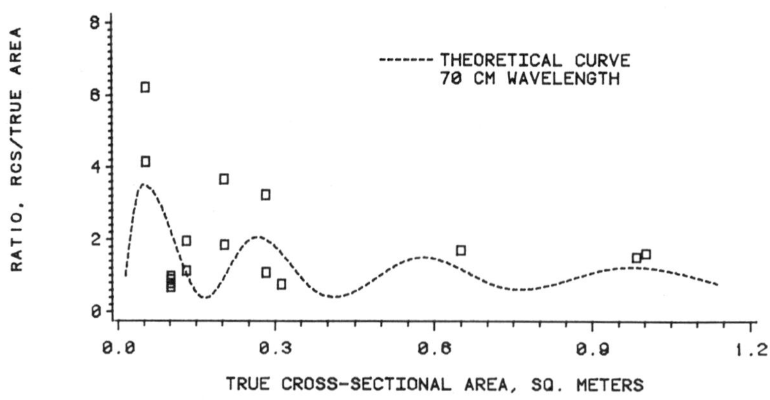

Fig. 1 Ratio of radar (RCS) to true area for spherical satellites.

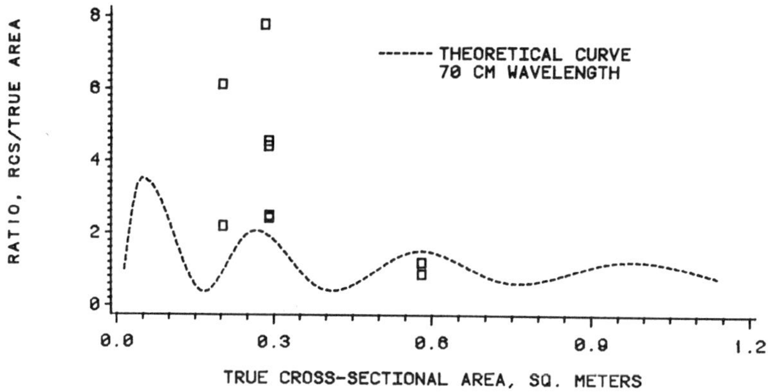

Fig. 2 Ratio of radar (RCS) to true area for "spheroid" satellites.

For this purpose, data for satellites of known sizes and shapes were used. First, the geometric cross-sectional area and RCS were compared for spherical calibration satellites—polished metal spheres of precisely known dimensions. The radar cross sections available for this comparison have all been measured using a radar wavelength of 70 cm. This wavelength is comparable to the dimensions of many of the calibration satellites. As a consequence, interference effects cause the RCS to vary sinusoidally as the geometric cross section changes. This effect is shown in Fig. 1, in which the ratio of RCS to geometric cross section is plotted against the geometric cross section, for both theory (dotted line) and observed ratios (data points). The data points follow the theoretical curve approximately, but with a good deal of scatter. On average, the observed ratios differ from the theoretical by about 20%. However, the standard deviation is close to 100%.

Another similar test was done using "spheroidal" satellites, that is, satellites which were approximately spherical, but may have had flat surfaces, or may have been ellipsoidal. These results are plotted in Fig. 2. The observed ratios fall well above the predicted values (particularly for the smaller satellites), averaging about a factor of two larger. These results suggest that irregular satellites have significantly larger RCS than geometric cross section. The data sample is very small, so this must be a tentative conclusion; but this result is consistent with what would be expected from the optics of a diffusely scattering object, as compared with a specularly reflecting object.

Recall that radar cross sections are always measured at zero scattering angle because of the geometry of the measurement technique. By general agreement, the RCS assumes a specularly reflecting sphere. Figure 3 shows a comparison of the intensity of scattered light from a diffusely reflecting sphere with that from a specularly reflecting sphere. At zero phase angle, a diffusely reflecting sphere will return about 2.6 times as much light as a specularly reflecting one. The same relation must hold true for reflection of radar waves. Consequently, for cases where the object reflects radar signals

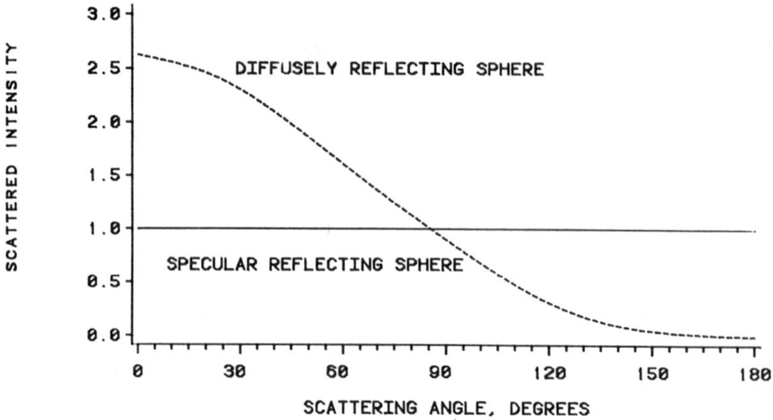

Fig. 3 Reflected intensity vs scattering angle for polished and diffusely reflecting spheres.

as a diffuse reflector, the RCS calculated on the assumption that the object is a specular reflector will be larger than the geometric cross section.

We see evidence that RCS reported in the Satellite Catalog may, on average, be larger than the geometric cross section by a factor of 2, or possibly more.

The factor "larger" is not a simple constant value, but depends on the shape and material of the debris and the way that radar data from successive measurements have been averaged. Unfortunately, in the absence of any other information, we had no choice but to use the nominal RCS for our albedo calculations, realizing that the values of "albedos" calculated from

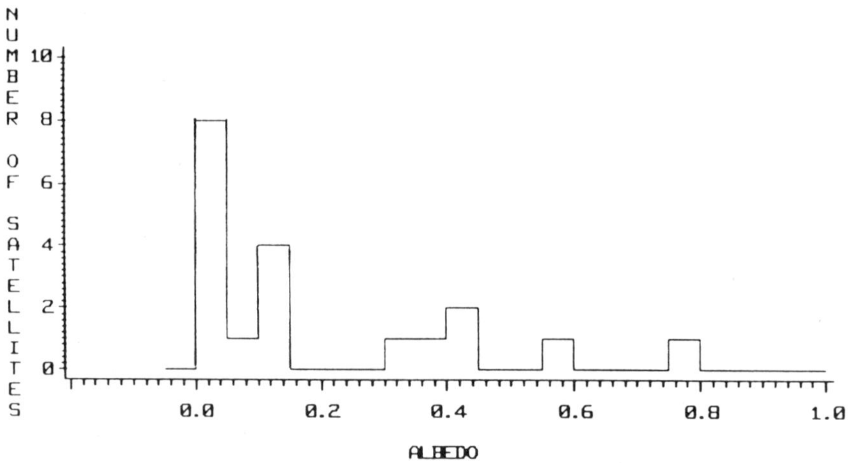

Fig. 4 Distribution of albedo for Ariane debris, observed and "not seen" objects.

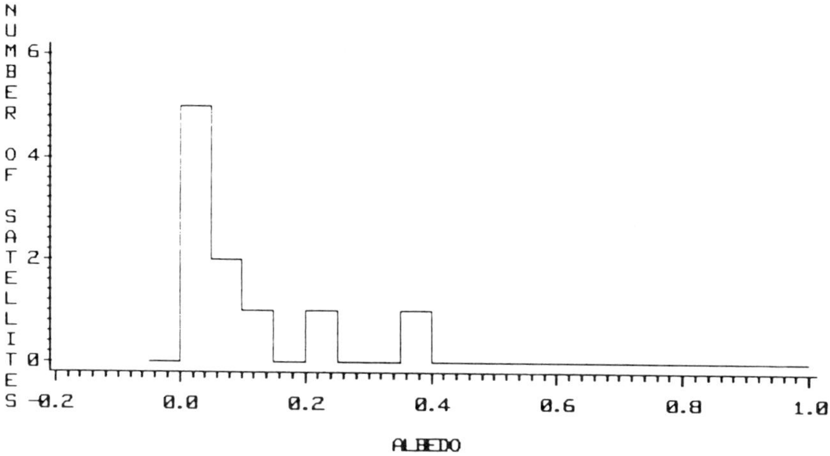

Fig. 5 Distribution of albedo for Landsat 1 debris, observed and "not seen" objects.

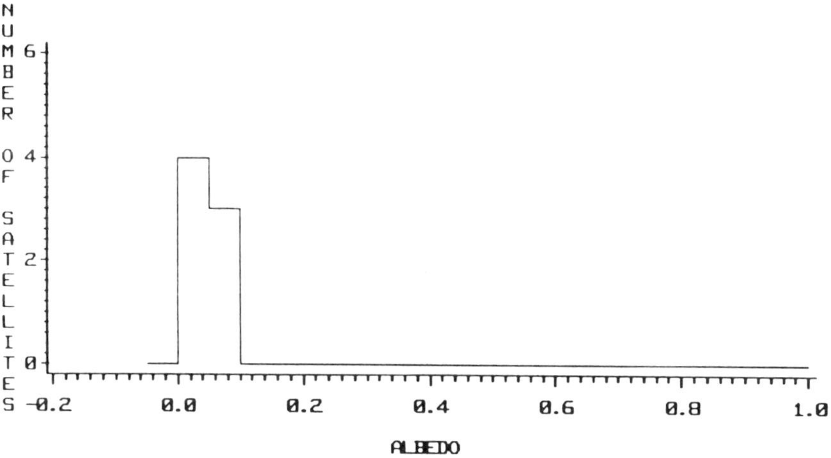

Fig. 6 Distribution of albedo for Landsat 3 debris, observed and "not seen" objects.

the RCS are probably too small by a factor of two or more. In all the following, the albedos reported have been calculated using the RCS and are thus subject to the foregoing qualification.

Results

Figures 4–8 show the number density of albedos observed for the five debris swarms measured in this research. Albedos from satellites "not

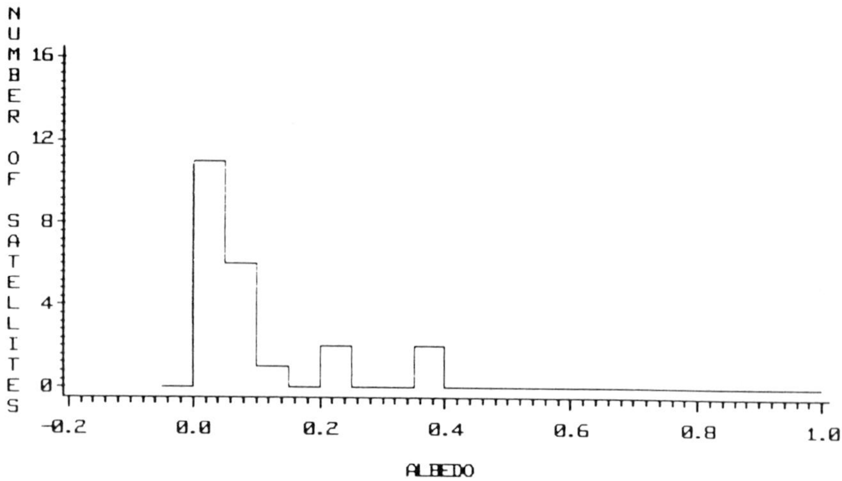

Fig. 7 Distribution of albedo for Cosmos 1275 debris, observed and "not seen" objects.

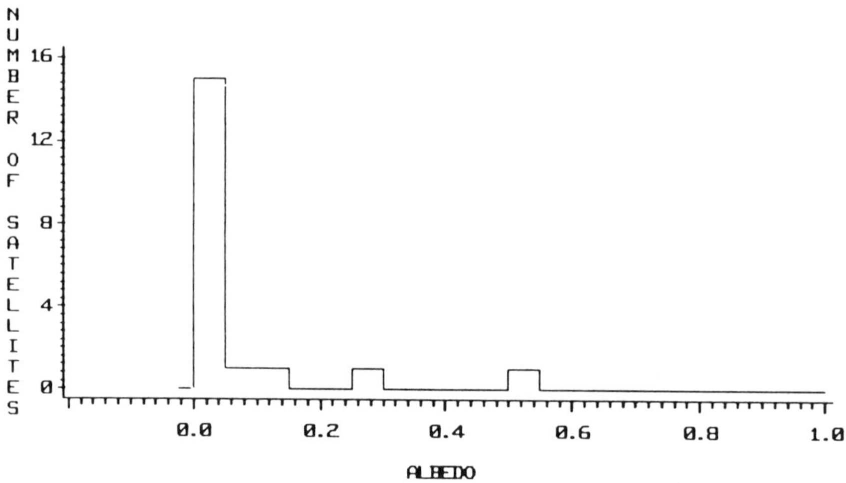

Fig. 8 Distribution of albedo for Solwind debris, observed and "not seen" objects.

seen" are included in these figures. These values are for satellites that we did not detect, but are sure that they crossed the video telescope field of view. Since we know the minimum stellar magnitude which can be detected by the video telescope, we can put a lower bound on the magnitude of the satellite in question, and hence can calculate an upper limit for its albedo.

Debris from Solwind and Landsat 3 breakups are almost all very dark, with albedos less than 0.1. Debris from Cosmos 1275 and Landsat 1 are

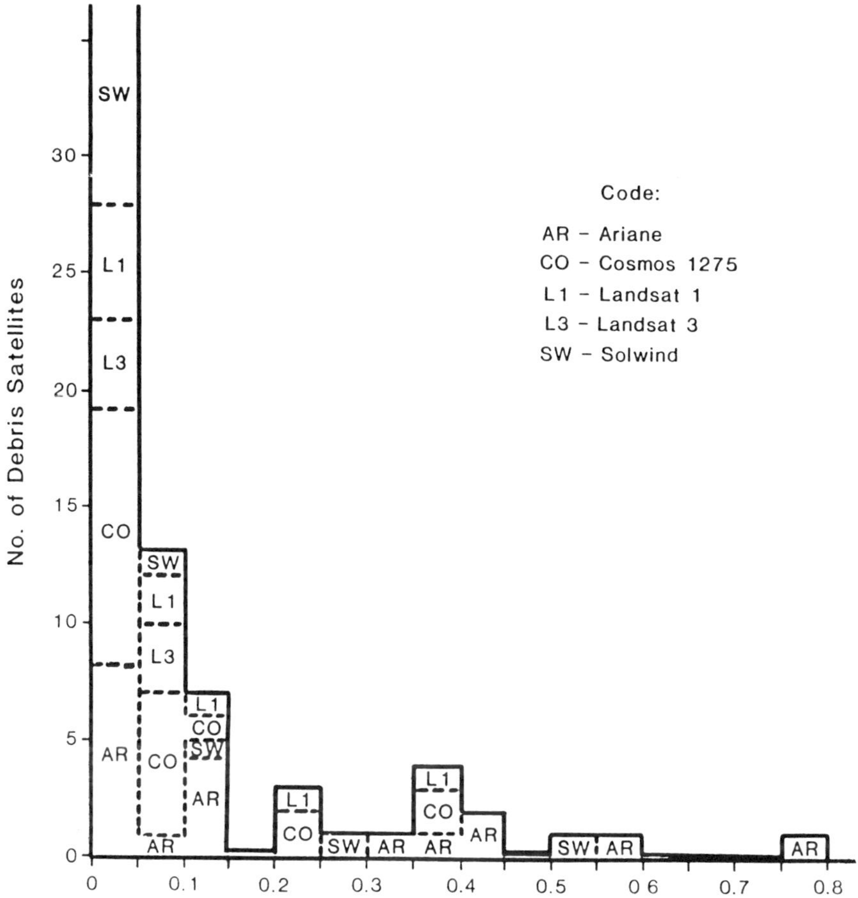

Fig. 9 Summary of all measurements: observed and "not seen" objects.

Table 2 Mean albedos of debris fragments from various breakups

Breakup	Probable cause	Mean albedo[a]	Standard deviation
Ariane	Unknown	0.196	0.231
Landsat 1	Propellant explosion	0.103	0.119
Cosmos 1275	Unknown	0.087	0.112
Solwind	Hypervelocity impact	0.065	0.124
Landsat 3	Propellant explosion	0.042	0.231

[a]based on RCS.

mostly dark, but with a few objects reflecting almost half the light they receive. The debris from Ariane include some very dark objects, but in general the Ariane debris are relatively bright.

The summary histogram of Fig. 9 shows albedos from all 75 satellites measured. The major peak occurs at the lowest albedo range, for which 43 values less than 0.05 were found. A low peak in the range of 0.4 to 0.4 has 14 objects. More than half of the "bright" objects are provided by the Ariane swarm.

The data are summarized in another format in Table 2, which gives the mean albedo and standard deviation for each breakup. Standard deviations are generally larger than the mean value. Ariane provides the brightest fragments. The debris from all the other breakups are relatively dark, with albedos of 0.1 or less.

Even taking into account that the "albedos" presented here are probably a factor of 2 or more too small (owing to uncorrected RCS), it is clear that most of the debris fragments reflect very little light. Albdeos of the order of 0.1 are the rule, rather than the exception.

Conclusions

We conclude that most debris fragments reflect light poorly, with albedos less then 0.1 (calculated, as explained, from RCS). Even if the true optical albedos are a factor of 2 or more larger, most debris fragments are very dark.

Moreover, we see no clear correlation between the cause of the breakup and the distribution and magnitude of the fragment albedos, although breakups do differ from one another in the distribution of albedo among the fragments.

Reference

[1]Nebolsine, P. E., Lord, G. W., and Legner, H. H., "Debris Characterization," Final Rept., PSI TR-399, prepared by Physical Sciences Inc. for Teledyne-Brown Engineering, Dec. 1983.

Chapter 8

Ariane-Related Debris in Deep-Space Orbit

L. G. Taff*
Lincoln Laboratory, Massachusetts Institute of Technology, Lexington, Massachusetts

Introduction

Several different optical searches for deep-space artificial satellites were conducted between October 1985 and June 1987. Each search concentrated on the equatorial region, and each used different telescopes, at different places on the Earth, for different durations, with different search rates, and with different limiting magnitudes. In addition to these efforts, the various radars used to track deep-space satellites, to maintain missile-warning fences, and to perform other surveillance functions frequently acquire deep-space objects of low-enough perigee. From the accumulated data it became clear that an extensive band of debris occupies the $n \cong 2$ rev/day, $e \cong 0.7$, $i \cong 7°$ orbital band. The 250-some new objects being routinely tracked (only 10 or so entered yet in the NORAD deep-space catalog) represent an exponential growth of material in this type of orbit.

This find added new point to the interest in deep-space satellite debris of several United States government agencies, such as the Air Force, which has responsibility to track all artificial satellites, and the National Aeronautics and Space Administration (NASA), which must plan protection of manned spacecraft.

The 250-some objects could not have been in orbit three years ago, for searches conducted then did not find them in statistically significant numbers.[1-3] Presumably these new objects derive from explosions of Ariane-related launch vehicles on July 2, 1985 and on March 18, 1986.

Optical Sensors

Descriptions of the observatories used for most of the search effort have already been published; this section will be brief.

The Experimental Test System (in New Mexico) of the Ground-based Electro-Optical Deep Space Surveillance (GEODSS) network was built by and is operated by the Space Surveillance Group of MIT Lincoln Laboratory[4-6] primarily to search for and track satellites in deep-space orbit. Under

*Currently, Instrument Scientist for the Hubble Space Telescope Fine Guidance Sensors, Space Telescope Science Institute, Baltimore, Maryland.

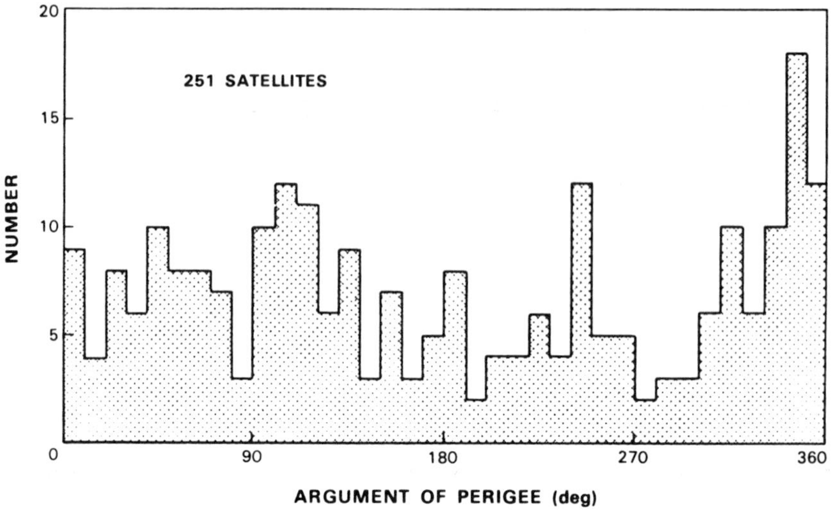

Fig. 1 Distribution of Ariane-related debris: inclinations; for 234 satellites.

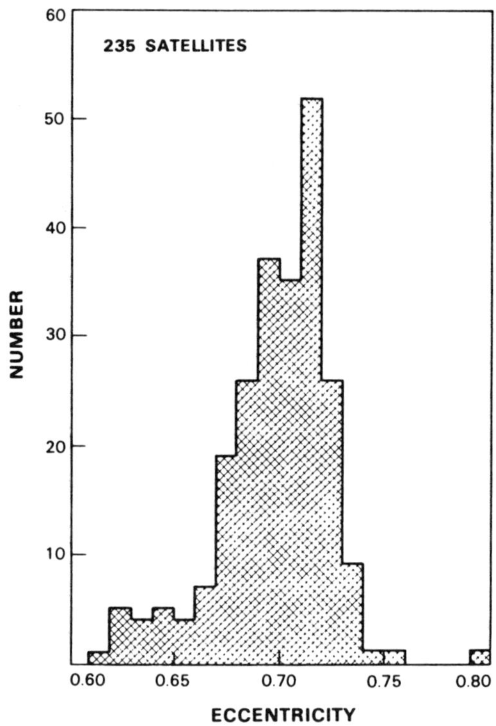

Fig. 2 Distribution of Ariane-related debris: longitude of the ascending nodes; for 251 satellites.

dark skies, and using video integration techniques, it will reach a limiting magnitude of $B \cong 17^m$ (the 50% probability of detection point). A poorly reflecting (albedo 0.1) piece of debris at opposition (this faint) would have an effective radius of $\cong 0.2$m at the geosynchronous distance.

After a proof-of-concept test in the winter of 1979 (the only deep, extensive optical search ever performed), the Air Force decided that a computer-controlled, electro-optically-equipped observatory (as the Experimental Test System was) could not only respond in real-time to tracking but would also complement the radar component of the Space Surveillance System. Hence, the Air Force contracted to have "replicas" of the observatory made and distributed worldwide. These formed the GEODSS system and are part of its Spacetrack system.

Descriptions of these observatories may be found in Beatty (1982).[7] Site locations include New Mexico, Hawaii, South Korea, Diego Garcia (all operational), and the eastern Atlantic (planned). In general, the GEODSS instrumentation is significantly less sensitive than its prototype (perhaps $B \cong 15^m5$), but has a search rate four times higher (600 sq deg/hr vs 150).

Over the 1986–1987 winter, all four GEODSS sites were used in an extensive, coordinated search. This effort also included the Baker-Nunn sites at St. Margarets, Canada, and San Vito, Italy.[8,9] A smaller-scale GEODSS/Baker-Nunn search in the winter of 1985 serves as part of the baseline for this report as does a NASA-sponsored search conducted at the

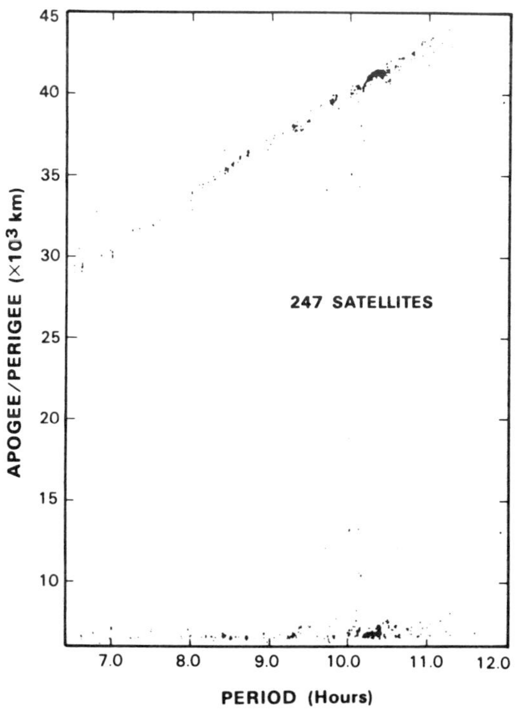

Fig. 3 Gabbard diagram of all Ariane-related debris: for 247 satellites.

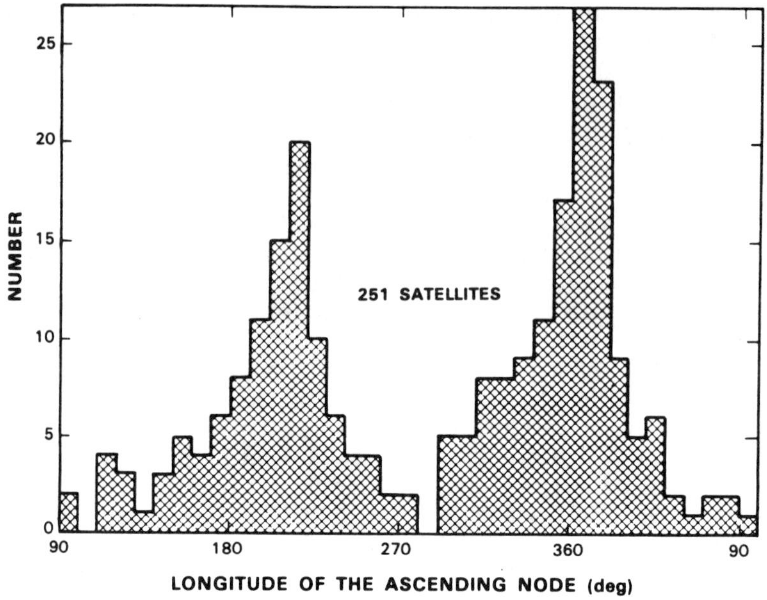

Fig. 4 Distribution of Ariane-related debris: eccentricities; for 235 satellites.

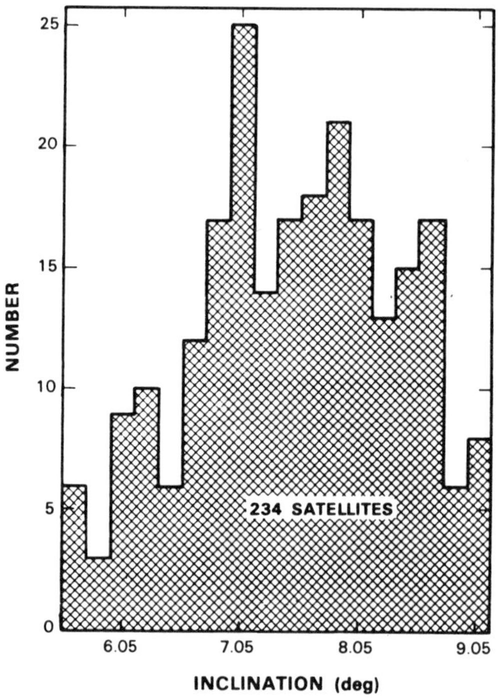

Fig. 5 Distribution of Ariane-related debris: arguments of perigee; for 251 satellites.

Experimental Test System in October 1985. Data from other optical searches have also been used in forming the conclusions of this report. Finally, a large number and variety of radars, such as the FPS-85 at Eglin Air Force Base (Florida) and the NAVSPASUR continental fence, have contributed to the database for the report.

Distributions

A combination of databases for deep-space artificial satellites have been searched for orbital element sets of the following properties: mean motion within 1.5 rev/day of 2.2 revolutions per day; eccentricity within 0.3 of 0.73; and inclination within 3 deg of 6.2 deg. These are the elements of a 1986 European Space Agency (ESA) rocket body. The inclination, longitude of the ascending node, and Gabbard diagrams for these objects are shown in Figs. 1–3.

The inclination distribution shows a broad, possibly multimodal form, centered near the Ariane launch latitude. The unshown wings go further southward than they do northward.

The longitude of the ascending node distribution shows a simple bimodal form with peaks on either side of the Ariane launch longitude.

A Gabbard diagram plots apogee and perigee height against period. Thus, each satellite appears twice. When one artificial satellite breaks up into many pieces, the Gabbard diagram of its associated debris has an oblique "X" structure. The arms cross (Fig. 3) at the original orbital period and height (a consequence of conservation of energy).

The eccentricity (Fig. 4) and argument of perigee (Fig. 5) distributions are less interesting. The nearly uniform aspect of the latter distribution is puzzling.

Summary

In summary, the result of a variety of optical searches for deep-space artificial satellites, when coupled with routine radar detection of deep-space satellites with low perigee heights, points to the rapid growth of objects in the Ariane-type orbit ($n \cong 2$ rev/day, $e \cong 0.7$, and $i \cong 7°$). These objects have appeared in a time frame consistent with two explosions of Ariane launch vehicles. Approximately 250 objects are now being tracked.

Acknowledgment

"The views expressed are those of the author and do not reflect the official policy or position of the U.S. government."

Supported by the National Aeronautics and Space Administration under contract NAS 9 - 17138.

References

[1]Taff, L. G., Beatty, D. E., Yakutis, A. J., and Randall, P. M. S., Advances in Space Research, Vol. 5, No. 2, 1985, pp. 35–45.

[2]Taff, L. G. and Jonuskis, D. M., Advances in Space Research, Vol. 6, No.7, 1986, pp. 131–137.

[3]Taff, L. G., Journal of Spacecraft and Rockets, Vol. 23, No. 3, 1986, pp. 342–346.

[4]Weber, R., Society of Photo-Optical Instrumentation Engineers, Vol. 203, 1979a, pp. 6–11.
[5]Weber, R., Optical Engineering, Vol. 18, 1979b, pp. 82–91.
[6]Taff, L. G., Publication of the Astronautical Society of the Pacific, Vol. 93, No. 555, 1981, pp. 658–660.
[7]Beatty, J. K., *Sky and Telescope,* Vol. 63, 1982, p. 469.
[8]Taff, L. G., "A GEODSS Search Exercise: Search Plan," Lincoln Laboratory, MIT, Project Rept. STK-145 I, 1987a.
[9]Taff, L. G., "A GEODSS Search Exercise: Results," Lincoln Laboratory, MIT, Project Rept. STK-145 II, 1987b.

Chapter 9

Potential Spot-1 R/B-Cosmos 1680 R/B Collision

Karl G. Henize*
NASA Johnson Space Center, Houston, Texas
and
Richard H. Rast†
Lockheed Engineering and Sciences Company, Houston, Texas

Shortly after the breakup of the Ariane third-stage rocket body (SSC No. 16615) on November 13, 1986, curiosity as to its cause led us to search the October 31, 1986 NORAD orbital-element catalog for other objects passing near the breakup site at about the same time. To our surprise, we found that the Cosmos 1680 rocket body (SSC No. 16012) passed through the same region with a height difference of only 1.8 km some 60 s before the time of the Ariane breakup. Since the probability that one of the roughly 6000 objects in the NORAD catalog passed through such a narrow space/time window is about .005, we undertook a search for orbital elements which would give more accurate data for the actual date of the breakup.

Detailed NORAD data revealed updated orbital elements for 16615 on November 1, and for 16012 on November 8. Figure 1 shows the resulting orbital ground tracks and object positions vs time. These are not significantly different from the ones derived from the October 31, orbital elements. Applying the maximum expected error due to the extrapolation of the orbital elements to the date of the collision and applying them in the most favorable direction, we find the smallest possible distance between the two objects to be 380 km. Therefore we conclude that a collision did not occur. The fact that a search for potential fragments of 16012 revealed no candidates whatever supports this conclusion.

The very low probability of such a conjunction so near the critical time leaves room for further speculation concerning a possible interaction between the two bodies. For example, the possibility that 16012 may be accompanied by several fragments too small to be detected by the NORAD radar system cannot be ruled out. If one of these had a slightly higher

*Senior Scientist, Space Science Branch.

†Principal Scientist, Solar System Exploration Dept.

apogee than the parent and thus lagged behind, it would have been in position to collide with the Ariane rocket body. This situation is also illustrated in Fig. 1. The dashed line shows the orbit plane of 16012 shifted westward by 1 min of Earth rotation. In this case, the intersection of the two orbits is shifted northward to a position roughly 100 km from the announced breakup point. Even such a speculative scenario does not lead to a convincing case for any interaction between the two bodies.

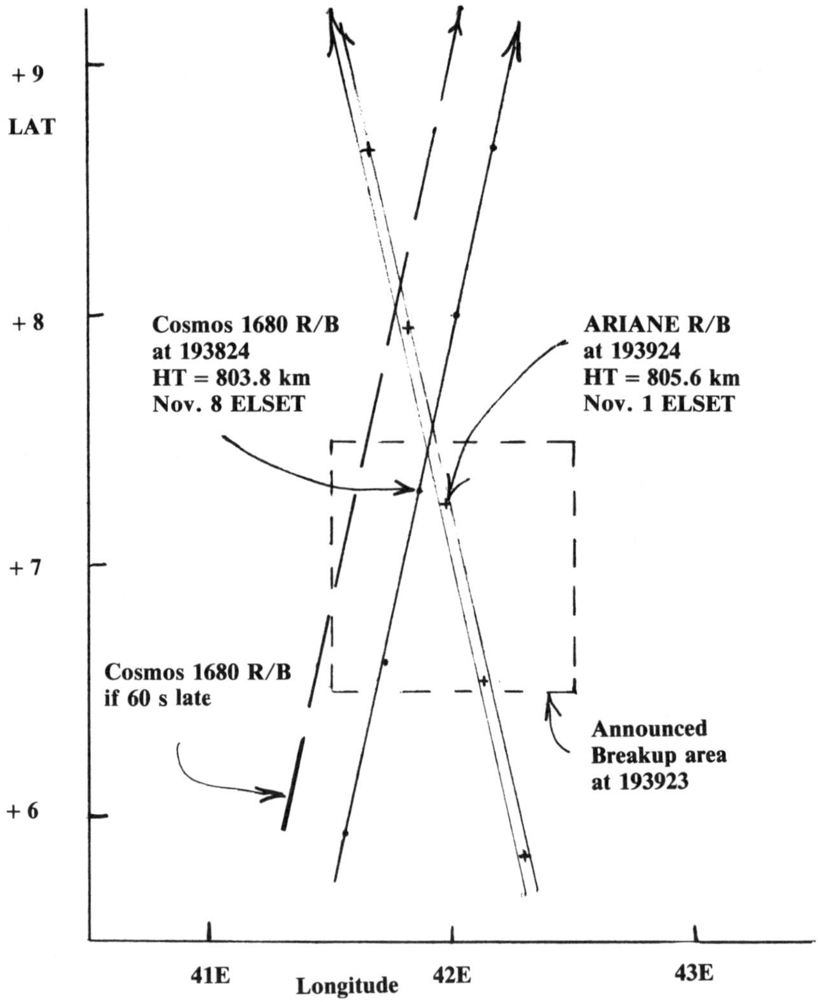

Fig. 1 Ground tracks for the Spot-1 R/B-Cosmos 1680 R/B near-miss. The announced breakup point was 42 deg east longitude, 7 deg north latitude. The size of the error box is an assumption based on the significant figures in the breakup-point position. Tick marks on the ground tracks indicate 12-s intervals.

It is interesting to note that, due to the near-equality of the orbital periods, a series of "near misses" took place between these two bodies at half-orbit intervals. Using the orbital elements noted above and making no allowance for propagation errors, we find that the two bodies passed within 58 km of each other 1.5 orbits before the time of breakup.

Control of Debris from Breakups

Chapter 10

Removal of Orbital Debris

Andrew J. Petro*
NASA Johnson Space Center, Houston, Texas
and
David L. Talent†
Lockheed Engineering and Sciences Company, Houston, Texas

Introduction

Man-made debris in Earth orbit—particles ranging in diameter from approximately 0.001 to 0.1 m—can by collision destroy a spacecraft. Enough objects now populate low Earth orbit (LEO) to make the likelihood of such collision non-negligible. Because much debris originates in the breakup of larger objects, means should be sought to eliminate such progenitors. This chapter describes techniques to remove debris objects already in orbit and operational procedures to constrain additions or remove debris in the future.

Any spacecraft in LEO may collide with a piece of debris. Total collision rate, with particles of size i and larger, will be as follows:

$$\dot{N}_i = (P_i)(A_c)(V_{\text{rel}}) \tag{1}$$

where $(P_i) = (\dot{N}_i / V)$ is the spatial density of objects of size (i) and larger in the volume V; A_c is the average cross-sectional area; and V_{rel} is the relative speed of the debris objects with respect to the spacecraft. The parameter A_c is determined by the geometry of the spacecraft, while V_{rel} is fundamentally dependent on the laws of orbital mechanics. Given these constraints, there

*Advanced Programs Office.
†Solar System Exploration Department.

is only one way to minimize the collision rate—by reducing the parameter P_i through the reduction of the number $\dot{N}_i$ of debris objects in orbit.

To appreciate the need for reducing debris, consider the effects of collisions with debris pieces of several different sizes.

It is convenient to divide orbital debris into three categories: 1) very small—typical diameter less than 0.001 m, 2) small—typical diameter of 0.001 to 0.1 m, and 3) large—radar detectable, and with diameter of 0.1 m and greater. Since mass increases as the cube of the radius (for spherical debris particles), the damage potential for objects in each of these categories can be described straightforwardly.

Assuming an impact speed of 10 km/s, solid aluminum spheres of 0.001-, 0.01-, and 0.1-m diam would deliver energy in the amount of 70.5, 70,500, and 70,500,000 Joules, respectively, upon collision with a larger object. Naturally, some of the energy would go into action other than the liquefaction of the area of impact (e.g., the ejection of fragments). However, *if* all of the available energy could be used to excavate a hole in the target (e.g., aluminum plate), hole depth would be about 0.004, 0.044, and 0.44 m, respectively, for the aluminum spheres just cited. Obviously, the hole depths for the 0.01- and 0.1-m cases exceed the thicknesses of aluminum plate found in most spacecraft—indeed, a complete breach of the plate and deposition of significant amounts of energy within the body of the target, and possibly its total disruption is implied.

Destructive potential can also be viewed in terms of the rate at which the energy would be deposited—the power. The energy would be delivered to the spacecraft in approximately the amount of time required for the impacting body to cross its own diameter. The average power during the impact for the three cases would be 705 Mw, 70.5 Gw, and 7.05 Tw, respectively. The average rate of energy delivery increases as the square of the radius of the particle (assuming the same impact velocity).

From the perspective of design, collision with debris has little practical concern if the typical time between collisions is much longer than the mission life. On the other hand, if the impact frequency by particles of a given size is very high, but the energy deposited per impact is very small, the practical concern is again low. The driving concerns will be *both* a significant chance of a collision *and* non-negligible energy deposition upon collision.

Present-day estimates suggest that the debris flux for particles of 0.0001-m diam and larger might produce about 10 impacts per cross-sectional meter squared (m^2) per year. Damage done by most of the impacts would be relatively small and easy to shield against. For objects having a diameter greater than or equal to 0.1 m, a collision may be anticipated only about once every 10,000 yr per cross-sectional m^2. A spacecraft having a cross-sectional surface area of 100 m^2 might experience an impact once per 100 years—a relatively low risk.

With these values setting boundaries on the problem, it becomes apparent that particles of characteristic dimension between about 0.001 and 0.1 m make up the critical range of debris threatening spacecraft of typical dimensions. Means have therefore been sought to remove both these small debris particles and the defunct larger objects that generate them, as well as reduce future production of orbital debris.

Removal of Small Orbital-Debris Particles

An object of large geometric cross section maintained in orbit for some time will sweep through the orbits of a predictable number of debris particles. The sweeper might retain particles or at least cause them to decay from orbit quickly. In either case, sweeper materials must sustain hypervelocity impacts without generating a significant amount of new debris; and the sweeper must otherwise be practical to build and deploy. This section describes the operation of debris sweepers and the practicality of using them. For ease of calculation and comparison, all debris objects will be assumed to be aluminum spheres.

A spherical balloon filled with some type of foam could retain particles it swept up. Under hypervelocity impact, a colliding particle would probably disintegrate, its constituents being captured.

The foam-ball sweeper, necessarily of large cross-sectional area, would undergo atmospheric drag. Some way would have to be chosen to nullify this. Or, a circular-plate sweeper could replace the ball. Oriented edge-on to the velocity vector, the plate would present a small cross-sectional area to the residual atmosphere, but enough surface area for oblique impact of a large amount of debris. The sweeper would not capture particles, but rather would reduce their time in orbit. That is, small debris particles, as they pass through the sweeper surface, would experience a momentary impulse that would reduce the major axis of their orbits, place them in a region of higher drag, and thus, reduce their time on orbit. A change in orbital speed of only 1% at 500 km can reduce the time on orbit by about 98% for a spherical aluminum particle having a diameter of 0.01 m; at 900 km, for the same delta-V of 1%, the reduction would be 93%.

For this planar debris-sweeper, consider a volume of space in LEO defined as a shell between 500 and 600 km, for example. Further, suppose a large number of debris particles exist within this volume, randomly distributed among all possible orbits subject to the inner and outer limits of the shell. Into an orbit that can range between the high and low bounds of the shell, introduce a planar debris sweeper; and to minimize the rate of decay of its orbit due to drag, orient the sweeper such that the unit normal vector to the plane of the sweeper surface becomes a right angle to its velocity vector. In keeping with the concept of a simple passive system, the sweeper would take no action either to intercept or avoid an object; the sweeping would be entirely *blind*.

The real debris population behaves quite complexly. To model the LEO environment as a series of concentric shells, it would be necessary to take into account a variety of factors influencing the population in each shell, including drag, collisions, explosions, debris objects that spend some time in several shells, deposition of new material on orbit, and removal of material from the environment. The model would be a system of coupled, nonlinear partial differential equations that would be tractable only by numerical methods. However, to investigate the effect of a single debris sweeper, free to interact with a population of identical particles confined to a closed shell, we may rewrite Eq. (1) as follows:

$$-\,\mathrm{d}N = LN\,\mathrm{d}t \tag{2}$$

where $-\,\mathrm{d}N$ is the number of particles that decay in time $\mathrm{d}t$ due to interaction with the sweeper and L is the decay constant, the probability per unit time of decay of any randomly selected debris object.

The solution of Eq. (2) yields a simple exponential decay:

$$N = N_0\, e^{-Lt} \tag{3}$$

The rate of the decay is determined by L, the decay constant, which by comparison to Eq. (1), is :

$$L = (1/V)(A_c)(V_{\mathrm{rel}})T_{yr} \tag{4}$$

where

- V = volume of shell = $4/3\pi(R_2^3 - R_1^3)$
- A_c = sweeper/particle interaction cross section; very nearly equal to πR_s^2 in most cases—R_s being the sweeper radius.
- V_{rel} = relative speed of sweeper to colliding objects; approximately equal to V_c, circulate-orbit speed, times $\sqrt{2}$.
- T_{yr} = number of seconds in a year to express time constant in terms of yr^{-1}.

For the case being considered here, using values appropriate for the 100-km-thick shell extending from 500 to 600 km:

$$L = 3.97 \times 10^{-9}\, A_c\ \mathrm{yr}^{-1} \tag{5}$$

an explicit expression for the decay constant, as it depends on the *total* sweeper area present in the shell.

One approach to determining the total sweeper area required would be to specify the desired half-life for removal of the population. From Eq. (3) the half-life is:

$$t_{1/2} = (0.693/L)\ \mathrm{yr} \tag{6}$$

If the objective was to reduce the population of debris objects in orbit by 50% in 5 years, the required value of L would be 0.139 yr^{-1}, which yields, from Eq. (5), a value for the total sweeper area of $3.49 \times 10^7\ \mathrm{m}^2$—implying a radius of 3333 m.

It would not be likely that such a large circular sweeper would be constructed; but the same overall area could be achieved by deploying 100 smaller sweepers, each with a radius of 333 m. Whether or not such a system of sweepers would be practical to deploy would be strongly related to their total mass.

The sweeper mass characteristics may be described in terms of a mass per unit area, ρd (kg/m^2), required to produce a chosen delta-V for debris particles of a particular radius and mass. If the collision of a debris particle with a sweeper can be treated as a perfectly inelastic collision, the surface mass density may be calculated as follows:

$$\rho d = [m_p(\Delta V/V_i)]/[A_d(1 - \Delta V/V_i)] \text{ kg/m}^2 \tag{7}$$

where

ρd = product of planar sweeper density and thickness of the plate.
m_p = mass of the debris particle.
A_d = cross-sectional area of debris article.
$\Delta V/V_i$ = percentage change required in velocity.

Using Eq. (7) and designing for a 1% delta-V for 0.01-m- diam debris objects yields a ρd value of 0.19 kg/m^2. Thus each of the 100 planar sweepers would have a mass of 66,300 kg. The total sweeper mass for the shell would be 6,630,000 kg. Expanding the effort to include all of LEO from 500–1500 km, the total sweeper mass would grow to 66,300,000 kg. (As a point of contrast, the total mass of man-made materials in orbit is about 2,000,000–3,000,000 kg.) Thus such a system of blind planar sweepers would require desposition in orbit of about 20–30 times the amount of mass already there. Furthermore, assuming the development of a Heavy Lift Vehicle (HLV) capable of placing 90,000 kg in LEO, roughly 750 launches would be required to deploy the entire sweeper system.

Obviously, such a system would be very difficult to justify. In addition, assuming that the sweepers would operate blind, they would sweep up valuable payloads as well as debris. When 90% of the debris was gone, 90% of the nondebris objects would be gone as well. Moreover, such a system of blind sweepers would run a *very* high risk of collision with each other. Thus blind planar sweepers might well create far more debris than they would sweep up.

Any or all of the following modifications would tend to reduce the objections to the planar sweepers just described: 1) build in collision avoidance, 2) relax the sweep-out time, 3) restrict sweeping to selected altitude bands of particularly high interest, 4) use sweeping for maintenance of the LEO-environment status quo instead of overall cleanup, and 5) allow less-massive particles and/or smaller delta-V to drive the mass characteristics of the sweeper.

An alternative to a large sweeper would be a "smart" sweeper—a relatively small device that would achieve a large sweep radius by virtue of sophisticated on-board sensors to detect debris particles. Active systems would transfer energy or momentum to the particles to accelerate decay from orbit. Such relatively autonomous spacecraft could use concentrated dust clouds, particle beams, or laser beams to decelerate particles.

Smart sweepers could also be designed to avoid active payloads. The sweeper's "kill radius" would be determined by the degree of sophistication of the sensor and fire-control system. Ideally, a dozen or so smart sweepers with a kill radius of about 3.5 km could replace the entire system of physically large sweepers. The mass of each of these units should be far less than the 66,300 kg estimated for the blind planar sweeper.

Removal of Large Objects

Large objects should be retrieved from orbit or forced into early re-entry before they can contribute significantly to the LEO debris environment.

An unmanned rendezvousing spacecraft might remove large objects from the environment. The Orbital Maneuvering Vehicle (OMV) under development by NASA could serve this as well as many others functions. As an orbital-debris remover, three modes of operation have been identified for the OMV: direct deorbit, collection to an orbit where debris objects would be of less risk to critical payloads, and attachment of strap-on deorbit devices.

Cost would be an important factor in terms of OMV operational parameters. It has been determined that for objects in LEO, with mass less than 2000 kg, collection for later disposal costs less than direct deorbit. Figure 1

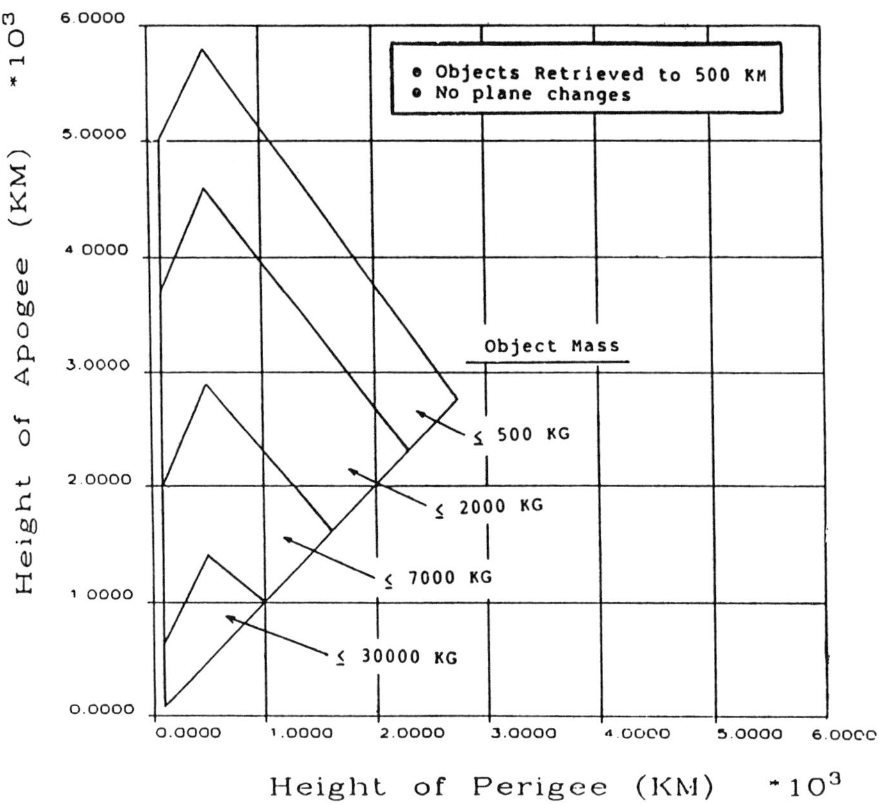

Fig. 1 Performance envelope of an Orbital Maneuvering Vehicle (OMV) as a debris remover. The OMV would be based in 500-km orbit.

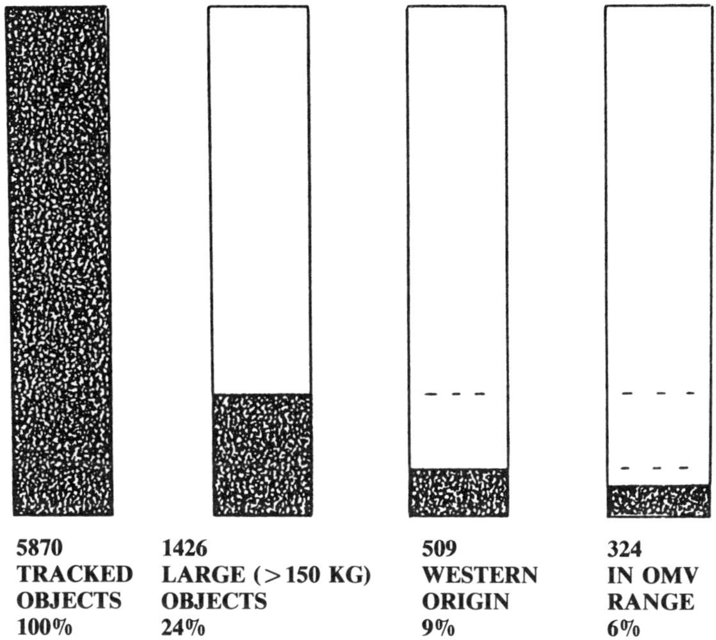

Fig. 2 Objects retrievable with an OMV; no plane changes. Western Origin means Western Europe, Canada, Japan, and the U.S.

shows the range of OMV operations as a function of apogee and perigee height as well as the mass of the debris object being retrieved.

As of mid-1987, 324 of 509 large debris objects (mass > 150 kg) of Western origin could be identified as falling within the OMV operational envelope (see Fig. 2). Furthermore, the majority of the objects were found to lie in a few narrow bands of orbital inclination (see Fig. 3). This is fortunate since an OMV based at a particular inclination could reach all of the objects at that inclination by simply waiting long enough for the orbital planes to align by natural precession.

Other than the direct deorbit or collection method, use of the OMV to attach a deorbit package to a piece of defunct hardware may be useful in some cases. Since the OMV propulsion system would not have to be used to change the orbit of the nonfunctional body, the performance envelope of the OMV would be expanded, bringing more large objects within its range.

Disposal Provisions for New Debris

For expendable hardware yet to be launched, premission installation of some type of deorbit device seems prudent. Two types of devices will be described—a passive drag-balloon, and a small propulsive deorbit system. Since most satellites have operational lifetimes measured in years, either type of system would have to remain inert safely for a long time and then function on command after other spacecraft systems had failed. And to be

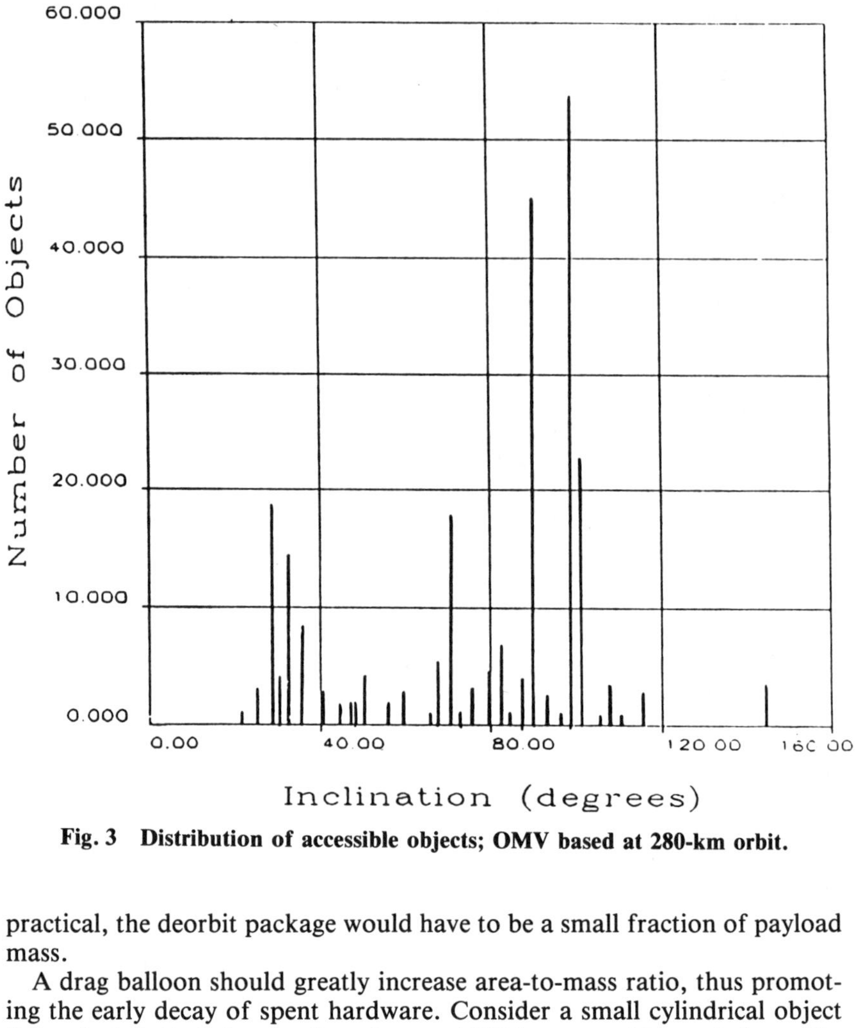

Fig. 3 Distribution of accessible objects; OMV based at 280-km orbit.

practical, the deorbit package would have to be a small fraction of payload mass.

A drag balloon should greatly increase area-to-mass ratio, thus promoting the early decay of spent hardware. Consider a small cylindrical object 6.4 m in height, 2.4 m in diameter, and 350-kg mass. Figure 4 shows the nominal decay time of this from a variety of initial circular orbits of increasing altitude. Time in orbit increases dramatically with altitude, from a few years at 500 km to nearly 1000 years at 1000 km.

For the debris cylinder defined above, altitude as a function of time was first calculated without the assistance of a drag device, and then, in succession, with drag balloons having radii of 2.5, 5.0, 10.0, and 20.0 m. All the drag devices are assumed to be balloons constructed of material having a surface mass density of 0.02 kg/m^2—similar to ECHO 1. Modern materials might be chosen that have a similar mass per unit area. This material was selected as a conservative example. Figure 5 shows the results of the calculations. Improvement between the "Without Device" case and the "5M" case is almost an order of magnitude in terms of the total time the debris object spends in orbit.

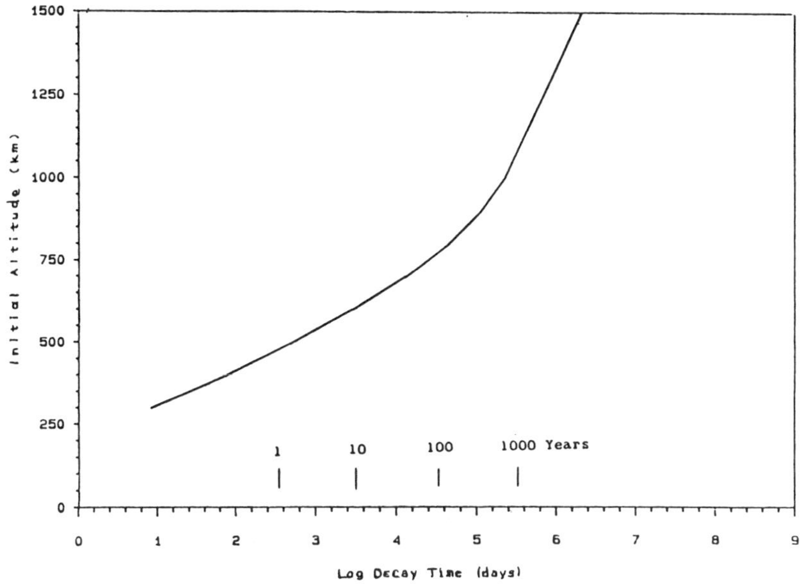

Fig. 4 Log decay time vs initial altitude.

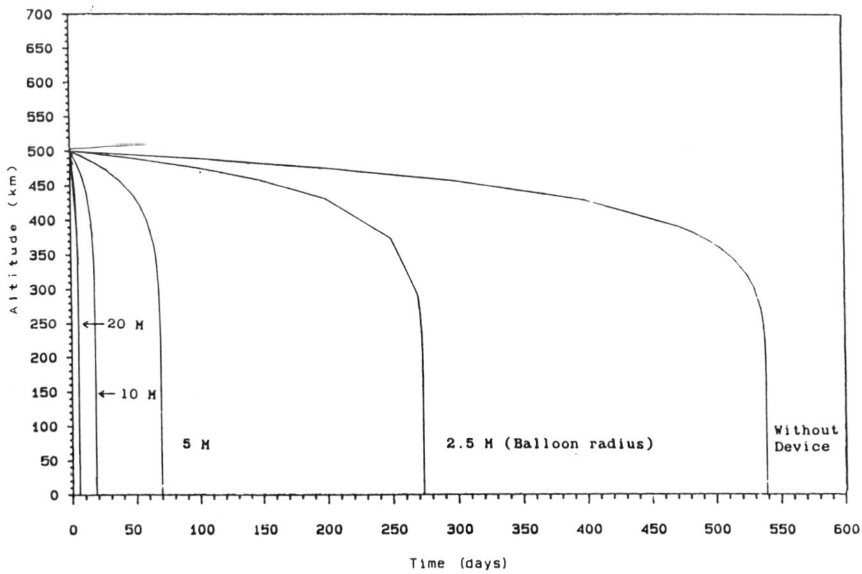

Fig. 5 Decay time for cylinder with and without balloon drag. Initial orbit, 500 km; 900-K atmosphere.

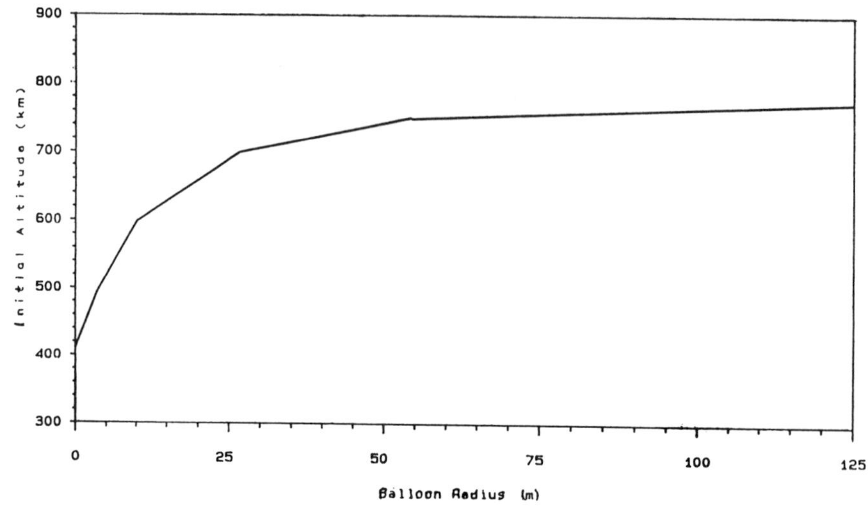

Fig. 6 Balloon radius required for 90-day decay as initial altitude increases. Balloon characteristics like Echo.

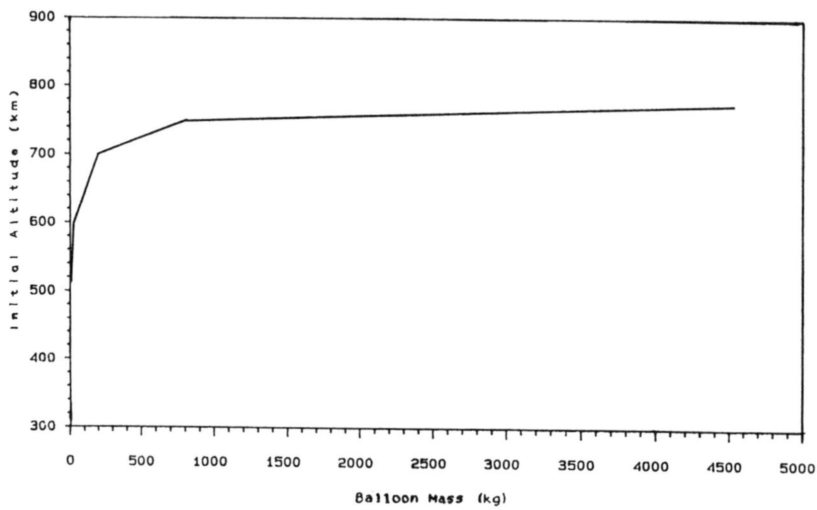

Fig. 7 Balloon mass required for 90-day decay as initial altitude increases. Balloon characteristics like Echo.

Since atmospheric density falls off rapidly with increasing altitude, there must be a ceiling for the use of balloon drag devices. To address this point, balloon radii and masses required to produce 90-day decays were calculated for the cylindrical debris object for successively greater altitudes. The results of these calculations (Figs. 6 and 7) indicate that, above about 750 km, passive drag devices are not practical. The practical band extends from, say, 250 to something over 500 km in altitude. This range might be extended to higher altitudes by using a device material having a surface mass density less than 0.02 kg/m^2. Also, the ceiling for practical operation of passive drag devices would be a function of the solar cycle.

For large debris objects that may be deposited at altitudes higher than 750 km, or as an alternative to drag devices altogether, small propulsive deorbit devices may be required. To illustrate a propulsive deorbit device, the cylinder, used for the drag calculations, will again be taken as the debris body. For all calculations the propulsion system was assumed to have a specific impulse (I_{sp}) of 310 s. Decay has been obtained when perigee drops below 80 km.

The results for a series of initially circular orbits are shown in Fig. 8 and summarized in Table 1. Of propellant-system mass, the fuel tank takes 5%;

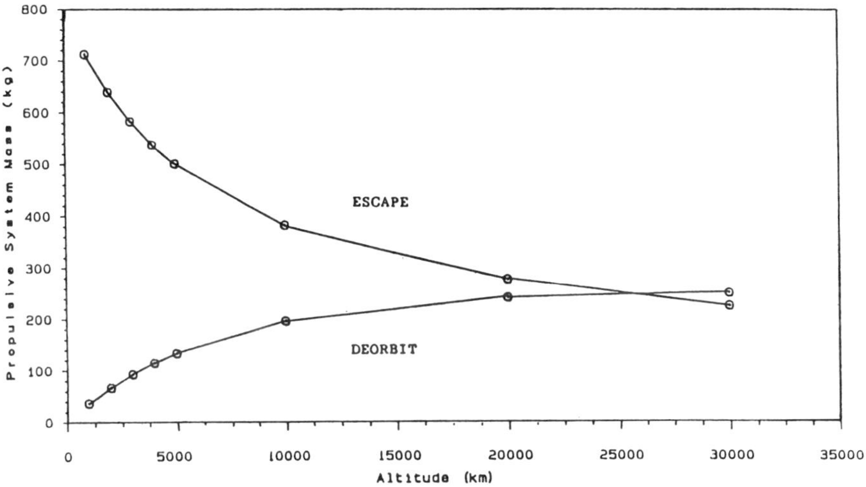

Fig. 8 Mass overhead for propulsive deorbit.

Table 1 Delta-V and propellant mass for deorbit to 80-km perigee or escape. Spacecraft mass, 350 kg; engine I_{SP}, 310 s

	Deorbit		Escape	
Initial orbit, km	Delta-V, m/s	Prop. Mass, kg	Delta-V, m/s	Prop. Mass, kg
500 × 500	121	15	3155	638
600 × 600	148	18	3132	630
700 × 700	174	21	3110	624
800 × 800	200	24	3088	617
900 × 900	224	27	3067	610
1000 × 1000	249	31	3046	603
1500 × 1500	361	44	2948	573
7500 × 7500	1088	150	2221	377
25000 × 25000[a]	1482	232	1477	231
35800 × 35800	1490	233	1274	191
35800 × 500	44	5	1671	256

[a]Crossover point for deorbit vs escape.

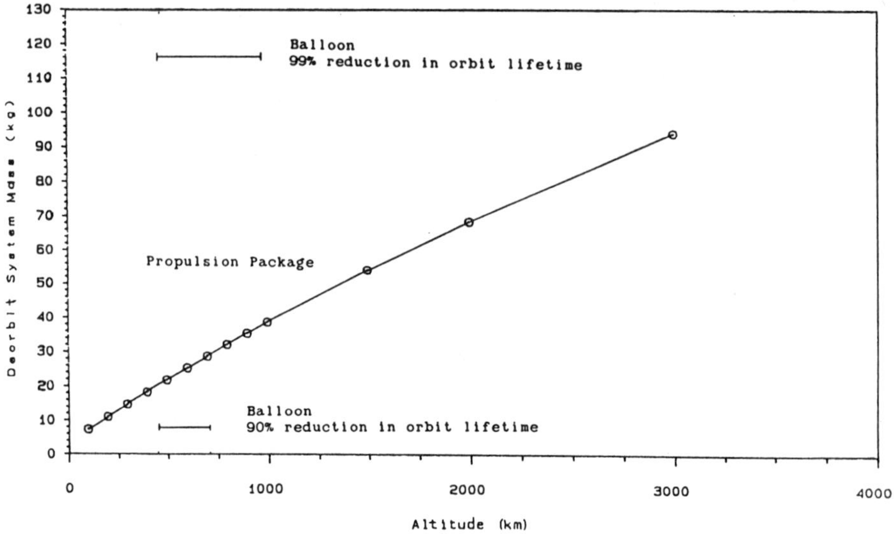

Fig. 9 Passive drag vs propulsive unit: mass overhead for deorbit.

the engine mass takes a fixed 7 kg. The graph's upper curve illustrates the mass of the propulsive system and fuel required to just escape Earth's gravitational field; the lower curve illustrates the propulsion-system mass required for re-entry. The crossover at 25,000 km implies that, for all lower altitudes, the cheapest (lowest mass-overhead) way to remove unwanted material from orbit will be to cause it to re-enter.

Above 25,000 km, the cheapest way to remove a debris object will be to place it in an escape trajectory. (As an aside, solar sails might possibly be used instead of a propulsive package.) However, this does not include the final disposition of spent geotransfer stages, which need only a small delta-V at apogee (about 50 m/s) to produce re-entry. In fact, for a geotransfer orbit having an initial perigee of 300 km, the entire propulsive-system mass would only be about 10–15 kg.

Furthermore, it can be seen from Fig. 8 that the propulsive-system mass, compared to the debris body, is not too unreasonable—less than 20% of the total out to about 1500 km. This suggests that propulsive deorbit packages may be particularly useful, with the important proviso that debris-body stability must be achieved before the deorbit burn. Control options include spin stabilization initiated by a pressurized gas jet, a simple Sun-sensor control system, a tractor rocket, and a tethered gravity-gradient rocket system.

With a properly executed propulsive deorbit burn, the debris object will re-enter almost immediately. In contrast, deployment of a passive balloon drag-device only diminishes time on orbit.

Figure 9 compares the deorbit system masses for propulsive and passive-drag systems (all calculations, again, for the cylinder debris-body). A 90% reduction in orbit lifetime can be achieved with less mass overhead by

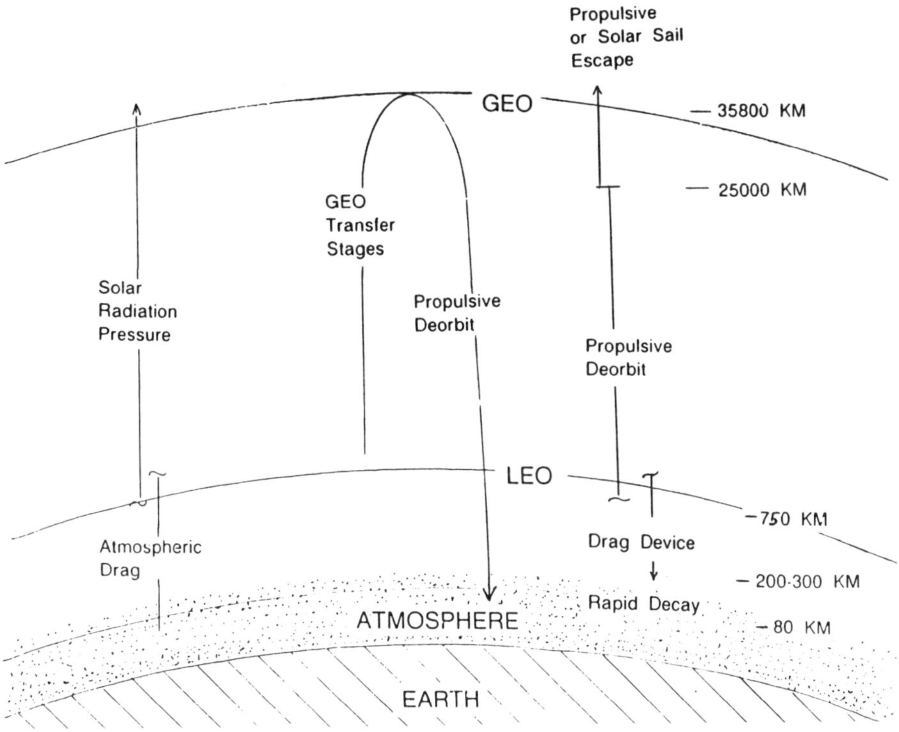

Fig. 10 Modes of debris removal and prevention by altitude band.

passive drag than by propulsive deorbit. But to reduce the time on orbit by 99% by the drag device would require roughly a factor of 4 times as much mass overhead as the propulsive system. (It should be noted that passive drag devices have a larger cross section than propulsive-deorbit devices, and thus a greater risk of collision with orbital debris.)

Summary and Conclusions

The several methods identified for diminishing orbital debris can be categorized as either preventative or remedial. Each technique for debris reduction or prevention has a regime of operation. The following list (illustrated in Fig. 10) summarizes modes of debris removal and prevention by altitude band.

1) *For all particles*:

250–2000-km altitude band: Smart sweepers will be the most probable type for objects in the 1 mm to several centimeter size-range. Large planar sweepers would probably need to be maneuverable to be practical.

2) *For large objects*:

80–250-km altitude band: No requirements for active removal are likely on a regular basis due to the short time for natural decay.

250–750-km altitude band: Use of deorbit devices indicated. Either a passive drag balloon or a propulsive deorbit package would be a viable option for coming payloads. Large objects already in orbit could be removed by an OMV in any one of its several operational modes.

750–2500-km altitude band: OMV rendezvous with the attachment of a propulsive deorbit package, or the OMV could grapple and dispatch old, large bodies, directly, as per the several identified options.

750–25,000-km altitude band: New, large bodies would require propulsive deorbit.

Over 25,000-km altitude: New, large bodies would require propulsive escape from Earth orbit.

Geotransfer stage: New, large bodies would require propulsive deorbit.

Early removal of spent flight hardware and nonfunctional payloads can reduce the rate of growth of debris in the orbital environment. Further study and development of "smart" sweeper strategies should address the problem of extant small debris. Use of an OMV, or its functional equivalent, can reduce the numbers of large debris objects, most of which lack means of self-disposal.

Chapter 11

Cost Estimates for Removal of Orbital Debris

Andrew Petro* and Howard Ashley†
NASA Johnson Space Center, Houston, Texas

Background

A one-year study was completed in 1987 on methods to control the hazards to spaceflight posed by man-made debris in Earth orbit. The large number of relatively small particles present the greatest hazard of collision with operational satellites. However, the study concluded that the most effective way to control the small-particle population would be to eliminate nonfunctional satellites and spent rocket stages from the orbital environment. These large collision targets can generate massive numbers of small but deadly particles.

There are currently no active measures taken to remove nonfunctional satellites or spent rocket stages from Earth orbit. These objects simply drift until atmospheric drag causes orbital decay. The study served to identify the following approaches for active removal of nonfunctional satellites and spent rocket stages: retrieval with an orbital Maneuvering Vehicle, deorbit by propulsive device, and deorbit by passive drag-augmentation device. Chapter 10 evaluates the technical feasibility of these techniques. Here we briefly summarize cost estimates for each approach. All costs are in 1988 dollars.

Retrieval with an Orbital Maneuvering Vehicle

Retrieval requires the use of a spacecraft similar to the proposed Orbital Maneuvering Vehicle (OMV) with modifications to extend mission duration and to add a satellite-recovery kit. The development cost for the recovery kit is estimated to be $21,000,000.[1] Each retrieval operation would usually require one dedicated OMV mission. OMV retrieval would be limited to objects in orbits no more than about 1000 n. mi. above the initial orbit of the OMV and within a few degrees of the OMV's orbital inclination. Retrieval of objects at high orbital inclinations would probably require the

Invited paper.

*Aerospace Engineer, Systems Analysis Branch, ED23.
†Program Analyst, Cost Analysis Office, BX24.

use of a ground-based OMV, operated from the Space Shuttle (STS). The user charges for the OMV would be as follows for each mission (1988 dollars):

Mission	OMV Operations	STS Operations
Ground-based OMV	\$3,700,000	\$17,000,000
Space-based OMV	\$3,100,000	\$12,700,000

Using the OMV to remove large objects at the same orbital inclination as the planned Space Station would cost approximately \$15,800,000 each using a space-based OMV. Removing objects at higher or lower inclinations would cost approximately \$20,700,000 each using a ground-based OMV.

Propulsive Deorbit Device

A propulsive deorbit device could remove spent rocket stages and short-lifetime satellites. It would be feasible for satellites at any orbital altitude or inclination. This approach, and the one that follows, are preventative measures. The devices would be part of the spacecraft equipment, at launch. The size and cost estimates for this approach assume that a propulsive deorbit device will be added to a geosynchronous transfer stage.

The deorbit device considered here would use storable liquid propellants with pressure-fed thrusters. The device might be rigidly attached to the rocket stage or deployed as a tractor rocket on the end of a short tether. The spent stage would have to remain relatively stable and be pointed in the proper direction for the deorbit maneuver. In the case of a geosynchronous transfer stage, the deorbit burn would occur at apogee, after the stage had separated from its payload. The mass of the device is estimated to be 70 lb.

Assume the following:

- Modifications are made to existing hardware and no new technology is needed.
- The launch vehicle is a Titan 34D and the upper stage is a Payload Assist Module (PAM).
- The deorbit device is sized to remove a geosynchronous transfer stage during its first orbit.
- The propulsion-system cost estimate is based on hardware proposed for the Space Station's Crew Emergency Return Vehicle.
- The attitude-control-system cost estimate is based on the average cost of control systems on ten typical satellites.

The payload mass must be reduced by an amount equal to the mass of the deorbit device (70 lb). In the case of an already-designed satellite, the weight penalty of the deorbit device could be reflected in the cost to reduce the weight of the satellite. A weight reduction of 70 lb was estimated to cost \$3,400,000 for engineering development and \$1,700,000 for each unit.

In the case of a satellite not yet designed, the weight penalty of the deorbit device can be indicated as simply an increase in the launch cost per pound of payload. This is how the weight penalty will be treated in the cost estimates for the propulsive deorbit and drag devices.

Launch-system cost would be as follows (1988 dollars):

Launch vehicle	46,000,000
Upper stage	8,000,000
Operations	28,000,000
Mission integration	21,000,000
Total	$103,000,000

The cost of the deorbit device would be as follows (1988 dollars):

Item	Development	Flight Hardware
Payload weight reduction (launch cost for 70-lb device)	—	$1,700,000
Propulsion system	$ 1,200,000	$ 400,000
Attitude-control system	$11,000,000	$3,200,000
Integration (83% of development; 37% of flight hardware)	$13,100,000	$1,900,000
Fee	$ 2,300,000	$ 600,000
Total	$27,600,000	$7,800,000

The addition of $7,800,000 to the $103,000,000 launch-system cost results in an increase of 7.5%. This does not include the $27,600,000 cost for design, development, and testing, which would be amortized over the total number of times that the deorbit device was used.

Passive Drag-Augmentation Device

The effect of atmospheric drag on a satellite can be increased by deploying a large balloon, i.e., increasing the effective area of the satellite without significantly increasing its mass. For objects orbiting below about 500 n. mi., a balloon with a diameter of 40 or 50 ft can reduce the orbital lifetime of the satellite from several years to several weeks. As one advantage of a drag device, the satellite does not need to maintain any specific orientation—needs no attitude control system. The balloon could be stored in a canister and be inflated after a rocket stage or satellite completed its mission. Further analysis is needed to determine if a drag device would be effective for a spent rocket stage in a geosynchronous transfer orbit, but this case was assumed for the purpose of comparison.

The drag device would consist of a balloon, canister, and inflation mechanism. The diameter required of the balloon depends on the initial altitude of the satellite and the desired reduction in orbital lifetime. The mass of the device would be almost directly proportional to the diameter of the balloon. For typical cases, the mass of the system would range, depending on application, between 200 and 600 lb.

Drag-device cost would be relatively insensitive to balloon diameter over the range being considered. Development cost for the drag device has been estimated to be $2,500,000. Flight hardware and integration of each unit would cost an estimated $500,000.

The payload weight penalty for the deorbit device would be reflected as an increase in launch cost per pound of payload. As the mass of the drag device ranges from 200 to 600 lb, the additional cost would range from $5,000,000 to $15,000,000. The total cost for the drag device would range from $5,500,000 to $15,500,000, or an increase of 5 to 15% over the basic launch cost. This does not include the $2,500,000 cost for design, development, and testing, which would be amortized over the total number of times that the drag device was used.

Summary and Conclusions

Recovery and removal of large objects with an OMV will cost between 15 and 20 million dollars for each object plus about 21 million dollars for development of the retrieval device (1988 dollars).

Providing a propulsive deorbit device for spent rocket stages will cost about 7.8 million dollars per vehicle, which is equivalent to a 7.5% increase in payload-delivery cost. Development of the deorbit device will cost approximately 27.6 million dollars. A lower cost for development and flight hardware would be possible if the rocket stage being disposed already had an attitude-control system.

Depending on the application, providing a drag device will cost between $5,500,000 and $15,500,000 per unit or an increase of 5 to 15% over the basic launch cost.

The preventative measure of providing a deorbit device for satellites and rocket stages appears to be more economical than the corrective measure of recovery with an OMV. The propulsive deorbit device should usually be less expensive and would have much wider application than the drag device. However, the low cost of the drag device makes it attractive as a option in certain specific applications, such as low-altitude satellites for which weight is not a critical design factor.

	Development	Cost per unit
OMV retrieval	$21,000,000	$15–20 M
Propulsive deorbit	$28,000,000	$7,800,000
Drag device	$ 2,500,000	$5,000,000–15,000,000

Reference

[1]"Concept Definition Study for Recovery of Tumbling Satellites," Martin Marietta Tech. Rept. MCR-86-1329, June 1986.

Chapter 12

Collision Probabilities in Geosynchronous Orbit and Techniques to Control the Environment

Bernell McCormick*
McDonnell Douglas Space Systems Company, Houston, Texas

This paper addresses the physical crowding problem of geosynchronous satellites by summarizing two independent analyses. The first analysis included developing a new technique for predicting the expected time between collisions of active geosynchronous satellites with expired geosynchronous satellites. The unique feature of this new technique is that deterministic methods are used to model the motion of satellites and statistical techniques are employed to estimate collision probability. This allows realistic distributions of active and expired satellites to be used in the prediction process. The results of this new technique compare very closely to the results of previously used techniques. The second analysis addresses disposal options for spent upper stages (PAM-D, IUS, etc.) that are currently left in stable elliptical orbits. These spent upper stages are a hazard to the geosynchronous region as well as low Earth orbits. Two propulsive techniques are presented that will reduce the orbit lifetime of the spent upper stages.

Introduction

The geosynchronous region of space, a finite natural resource, has seen both radio-frequency (rf) and physical crowding by satellites. This chapter addresses only the physical-crowding problem. Two independent analyses were performed to assist in defining the problem and to frame suggestions to overcome it. The first analysis included developing a new technique for predicting the expected time between collisions of an active with an expired geosynchronous satellite. This analysis will help define the potential for collisions and offer a database to aid international policy on the disposition of expired geosynchronous satellites.

The second analysis addressed the disposal options for spent upper stages (PAM-D, IUS, etc.) now being left in stable elliptical orbits. These spent upper stages have perigee altitudes of about 300 km and apogee altitudes of

*Group Engineer, Engineering Services.

about 35,800 km. They therefore present a hazard to the geosynchronous region as well as low Earth orbit (LEO). The analysis determined the orbit characteristics that would minimize orbital lifetime and two propulsive techniques for achieving them. These techniques will require hardware changes to an upper stage and will reduce the amount of payload that can be taken to geosynchronous orbit.

Collision Probability for Geosynchronous Satellites

Over a long period, perturbing accelerations will cause a geosynchronous satellite to drift from its initial Earth-fixed longitude. Figure 1 shows a very simple model of how the relative motion induced by perturbing accelerations creates the potential of an expired satellite colliding with an active one. The perturbing accelerations cause a periodic libration about stable Earth-fixed longitudes (minor axis of the Earth's equatorial ellipse) with periods of up to 5 years and periodic variations in inclination from 0 to 15 degs over a period of some 50 years. Active satellites employ station-keeping schemes to counteract the perturbing accelerations. But current practice for expired satellites is to let them remain and drift in the geosynchronous-orbit region—a traffic hazard for the actively controlled satellites.

Should this practice continue, the collision probability will become unacceptably high. A new technique has been developed to estimate the collision probability as a function of the number and distribution of expired and active satellites. This technique employs deterministic methods for modeling the motion of satellites and applies statistical theory to estimate collision probability. It has the advantage over previously developed techniques of exerting realistic distributions of expired and active satellites in the prediction process.

Techniques For Predicting Collision Probability

A spectrum of techniques for predicting collision probability has been employed by the satellite technical community, ranging from a purely statistical approach to a purely deterministic approach. The purely statistical approach—analogous to the kinetic theory of gases (random behavior)—offers no information about individual satellites, but does yield information about the *system* of satellites. This approach has been employed by Perek to offer a rough estimate of the collision probability.[1] At the opposite end of the spectrum, the deterministic approach employs state propagators to numerically integrate the equations of motion to predict the closest approach of two individual satellites. This approach requires very precise computation and accurate initial conditions. The Air Force Satellite Control Facility utilizes a deterministic method to monitor close-approaches and to plan satellite repositioning if a close-approach alert is issued.

An unrealistic amount of computer resources are required to model the hundreds of satellites in geosynchronous orbit. To reduce this computer burden, a new technique has been developed that uses state propagators to predict the motion of satellites and statistical techniques to estimate collision probability. It is assumed that the state propagator models the population of close approaches of satellites, but does not precisely predict close approaches of individual satellites. This relaxes the requirement for very

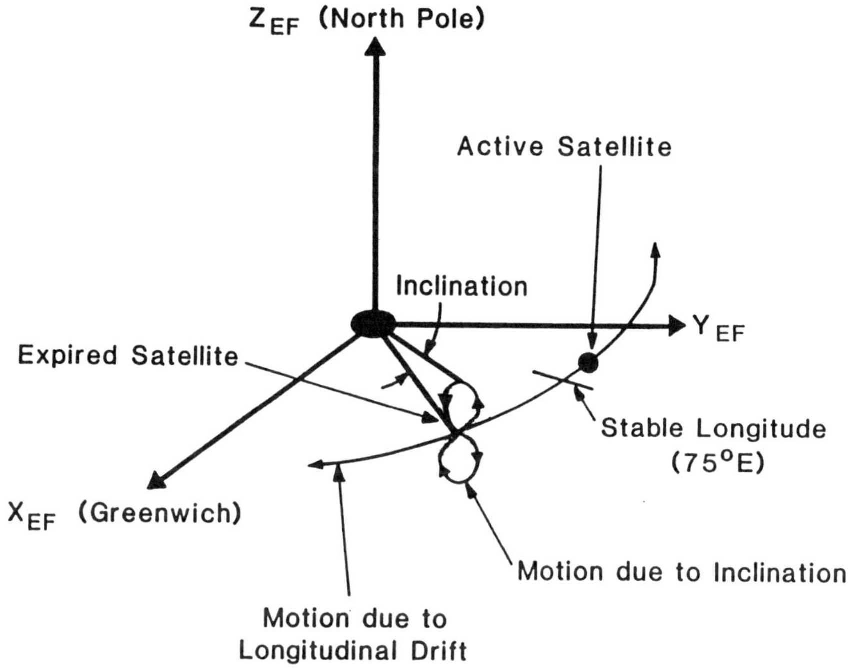

Fig. 1 Geosynchronous satellite motion.

precise state propagators needed for the purely deterministic approach. This new technique allows realistic distributions of expired and active satellites to be used in the prediction process.

New Technique

Basic approach

The new technique first determines the minimum approach distance between each expired satellite and the closest active satellite and then applies statistical techniques to extrapolate to collisional radii. This approach reduces the problem to one dimension (the minimum approach distance). The asymptotic theory of extremes suggests that the distribution of these minimum values can be represented by one of three limiting distributions.[2] The appropriate one for this problem is the two-parameter Weibull distribution, owing to geometric considerations (the distribution function of the Weibull approaches zero as the closest-approach distance approaches zero). (Applications of extreme-value distributions arise in analysis of flood flow, air pollution, reliability of complex systems, and breaking strength of materials.) The Weibull distribution function is represented mathematically as follows:

$$F(X) = 1 - \exp\left(-\left(\frac{X}{\beta}\right)^{\gamma}\right)$$

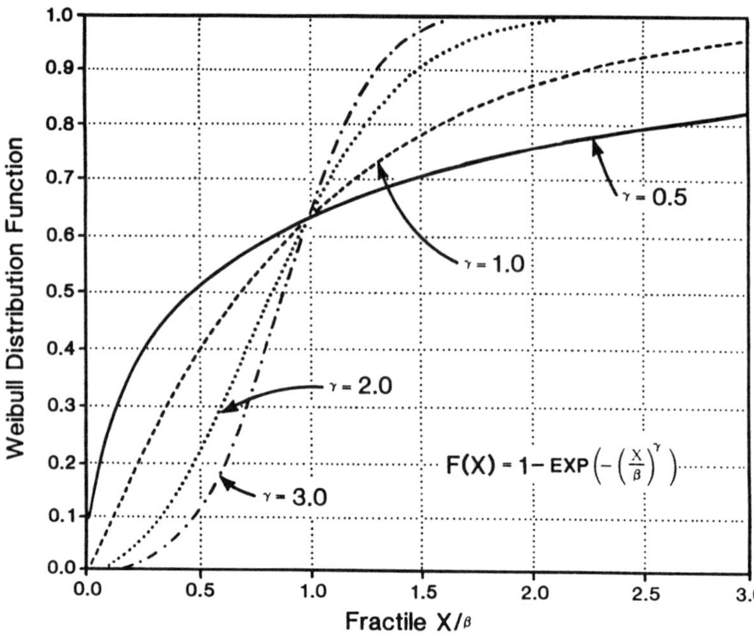

Fig. 2 Weibull distribution function.

where

$F(X)$ = accumulative probability
X = distance between satellites
β = scale parameter
γ = shape parameter

Figure 2 shows the Weibull distribution function vs fractile X/β for various values of the shape factor γ. This distribution gives the probability $F(X)$ of having a minimum approach distance of less than some distance X.

Minimum-distance data

An analytic state propagator (modified version of the NORAD propagator SDP4)[3] is used to propagate the entire population of expired satellites for an extensive period. The perturbations modeled by the propagator include Sun gravitation, Moon gravitation, and important Earth geopotential terms. The initial state conditions for the expired satellites are obtained from the NORAD satellite database. The initial conditions for the active satellites are based on a realistic model allowing for grouping at certain key longitudes. The state propagator is used to predict the minimum approach distance of each expired satellite to the closest active satellite. Since the active satellites tend to have near-zero inclinations, the minimum approach distance will occur at or near the equatorial plane. Perturbations will cause the expired satellites to cross the plane at a different place each transit. The active satellites will employ station-keeping schemes to counteract these perturbations—keep the satellite virtually stationary (in an Earth-fixed rotating coordinate system). Each expired satellite will have two minimum approach distances per day (two potential collisions per day). The analysis projects the course of expired satellites for years and thus portrays thousands of potential collisions (depending on the number of expired satellites and total propagation time).

In general, these minimum approach distances will be in the order of hundreds of kilometers; that does not give any information on collision probability. However, the asymptotic theory of extremes suggests that the distribution of these minimum approach distances can be represented by a Weibull distribution. A Weibull distribution function fitted to these minimum approach distances will allow for extrapolation to collision radii (meters). Two methods (method of moments and method of maximum likelihood) are employed to make the fit and give an estimate of the error in the fitted values of the shape parameter γ and the scale parameter β. Several goodness-of-fit test statistics are employed to determine how well the values of γ and β fit the actual data. These test statistics include the Anderson-Darling (A^2), Watson (U^2), and Cramer-von Mises (W^2) statistics.[4]

Extrapolating to collisional radii

The Weibull distribution gives the probability $F(X)$ of having a minimum distance X between an expired satellite and an active satellite. If the minimum distance becomes less than the average diameter of the two satellites, a collision will occur. The expected time between collisions can be determined as follows:

$$T_{\text{expected}} = \frac{1}{F(X)^{*}N}$$

where

X = average diameter of the satellites
$F(X)$ = probability of the minimum distance being less than X
N = total number of potential collisions per unit time

Numeric example

This new technique was applied to a configuration of 100 active and 54 expired satellites. The initial conditions for the inactive satellites were taken from the August 1983 NORAD satellite database. Figure 3 shows the expected time between collisions as a function of average satellite diameter. The expected time between collisions for a realistic average satellite diameter (~ 5 m) exceeds 10,000 years. A band of expected time between collisions is given for each value of satellite diameter. This band reflects uncertainty in the fitted values of the shape parameter γ and the scale parameter β of the Weibull distribution function.

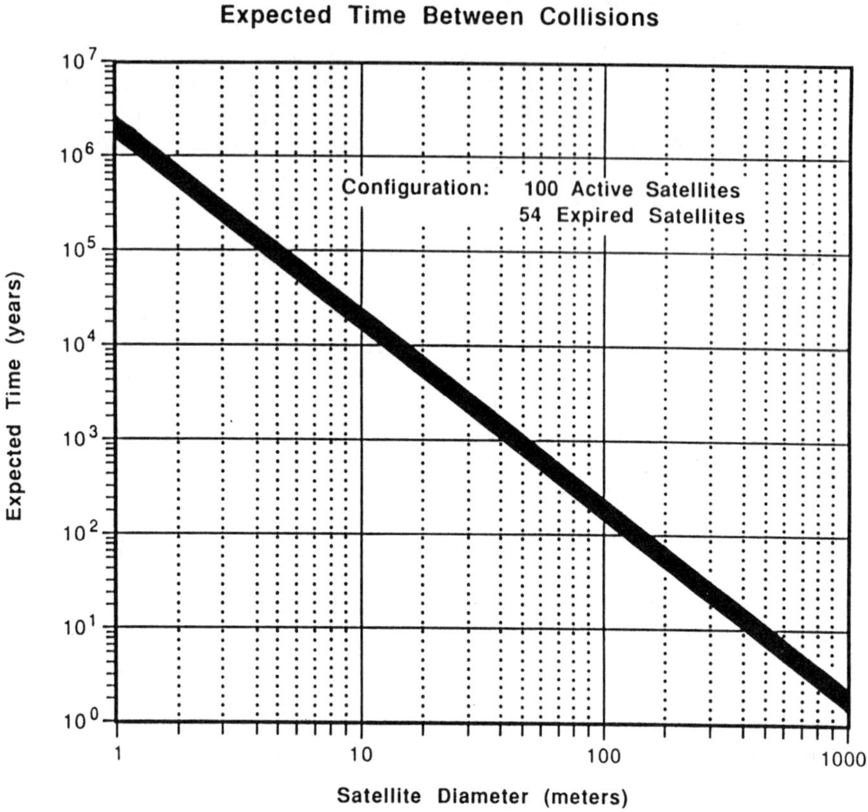

Fig. 3 Expected time between collisions.

Figure 4 compares results of the new technique to previous collision analysis. The average satellite diameter was assumed to be 50 m for each of the techniques. Perek's technique assumed uniform distributions of both the expired and active satellites. Hechler[5] and Chobotov[6] employ more realistic distributions of expired satellites, while assuming uniform distributions for active satellites. A more realistic distribution of active satellites takes into account the group positioning at certain key longitudes. The new technique was applied both to a uniform distribution and the realistic distribution of active satellites. The uniform distribution is used for comparison with other techniques; the realistic distribution offers a more reasonable estimate of collision probability.

These results suggest that the probability of collision is greater for the realistic distribution than was predicted for the uniform distribution.

Upper-Stage Disposal Options

Communication satellites brought to LEO by the Space Transportation System (STS) get boosted into geosynchronous orbit in two stages (Fig. 5).

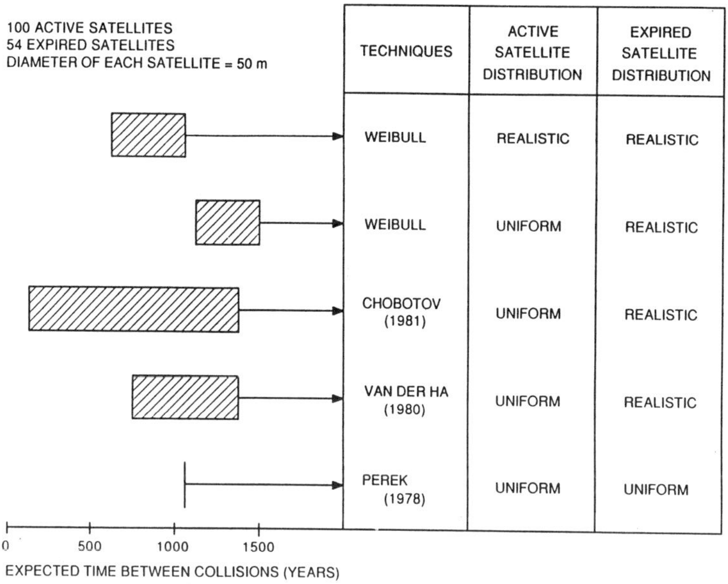

Fig. 4 Collision analysis techniques.

Fig. 5 Typical transfer orbit.

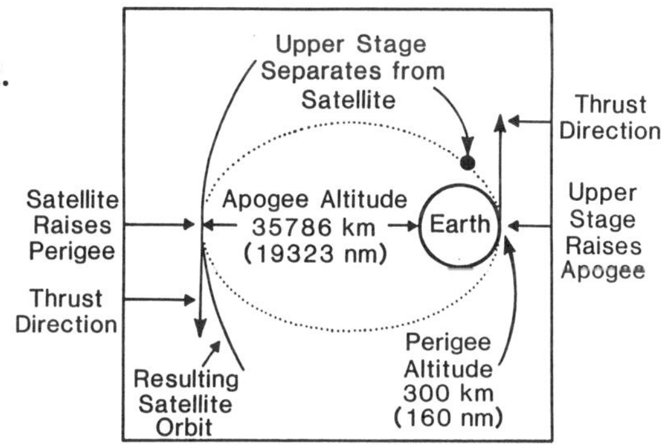

The satellite (after deployment from the payload bay) is inserted into an elliptical transfer orbit by an upper stage (PAM-D, IUS, etc.) and separated from it just after completion of the burn. Then the satellite will coast up to apogee, where a kick motor circularized its elliptical orbit.

The elliptical transfer orbit has an apogee of approximately 35,780 km and a perigee altitude of approximately 300 km. This orbit will have a stable lifetime of many years. A parametric study determined the desired orbit characteristics that would minimize orbit lifetime. Propulsive techniques were investigated to achieve the desired characteristics.

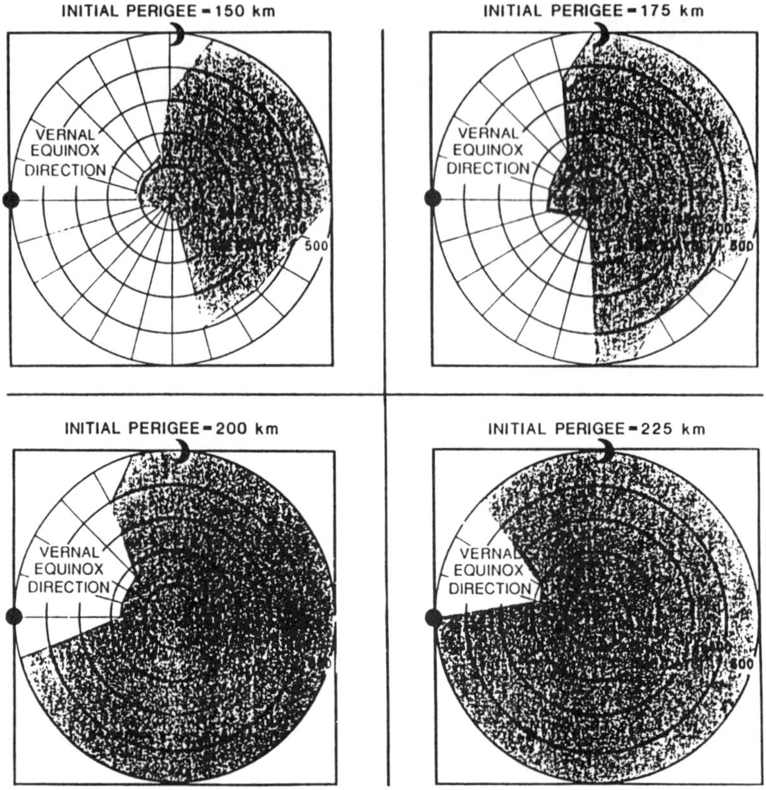

Fig. 6 Orbital lifetime in terms of initial perigee and initial apogee direction.

Parametric Study

The important perturbations affecting the lifetime of transfer orbits are Earth oblateness, gravity on Sun and Moon, and atmospheric drag.[7] A semi-analytic technique[8] that models these perturbations was used to study parametrically the initial perigee altitude, the initial apogee direction (with respect to the vernal equinox), and time of year of launch to determine effect on orbit lifetime. Figure 6 gives orbit lifetimes in terms of varying the perigee altitude from 150 to 225 km and varying the initial apogee direction with respect to the vernal equinox from 0 to 360 degs in 15-deg steps. The radial direction is the approximate time to re-enter, and the angular direction is the initial apogee direction.

This figure illustrates that pointing the apogee direction in the direction of the vernal equinox minimizes orbit lifetime. The window about this direction becomes larger as the initial perigee is decreased. The initial apogee direction can be controlled by choosing the appropriate launch time. As the initial perigee increases above 225 km, the orbit will have a lifetime of many years regardless of the initial apogee direction. Figure 7 shows the

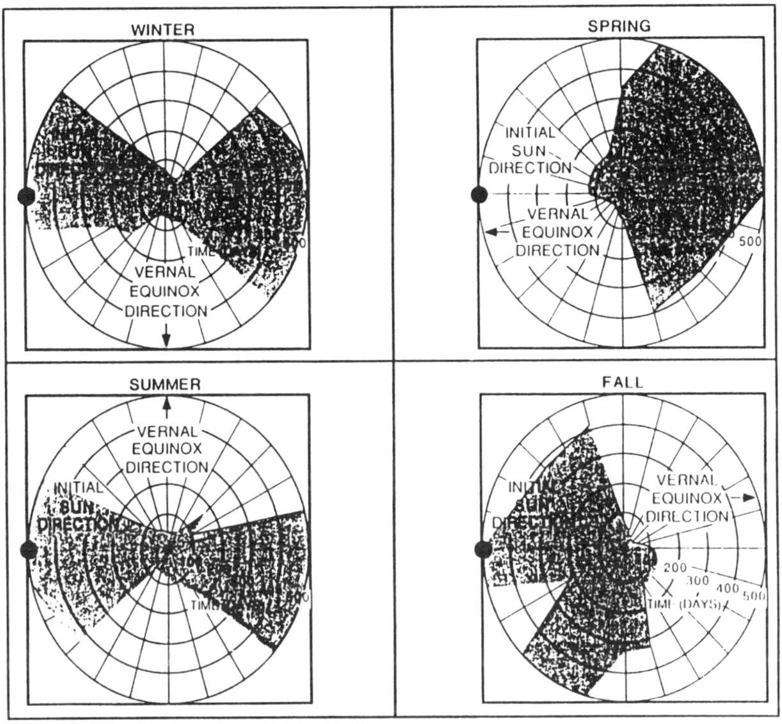

Fig. 7 Orbital lifetime in terms of time of year of launch and initial apogee direction. Initial perigee, 150 km.

Fig. 8 Geometry for propulsive orbit-changes.

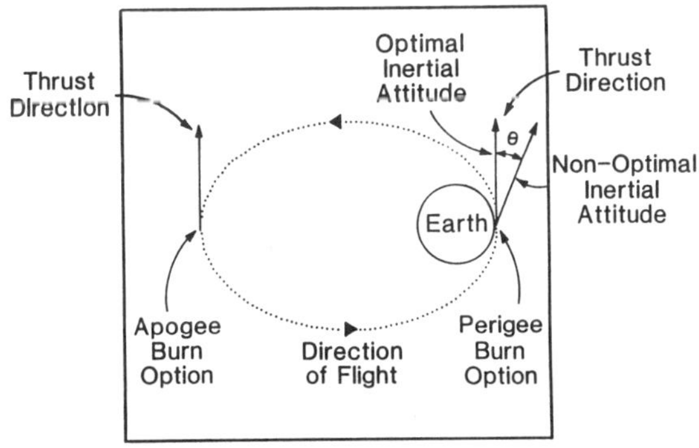

effect of time of year of launch on orbit lifetime (initial perigee, 150 km). This figure reinforces the idea of pointing the initial apogee direction at the vernal equinox to minimize orbit lifetime.

This parametric study emphasizes the need to lower the initial perigee altitude to minimize orbit lifetime.

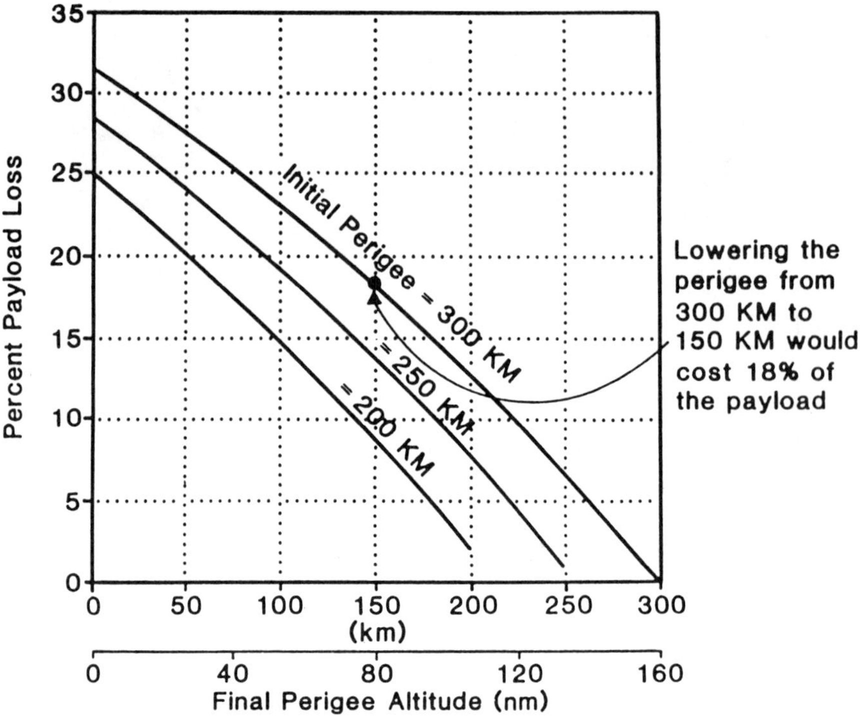

Fig. 9 Performance loss for lowering perigee using a perigee burn.

Propulsive Techniques for Lowering Perigee

Two propulsive techniques (Fig. 8) were investigated for lowering the initial perigee altitude: perigee-burn and apogee-burn.

The perigee-burn option will have a significant impact on the performance of the spent upper stage. It requires lowering the perigee with a burn near perigee, the least optimal place in the orbit (optimal being apogee). Because the upper stage will usually be loaded with the maximum amount of propellant, requiring nonoptimal performance lowers the payload that can be taken to geosynchronous orbit (Fig. 9).

The perigee-burn option could be used in conjunction with aligning the initial apogee direction in the direction of the vernal equinox. For an initial perigee altitude of 300 km, lowering the perigee cuts payload 18%. The parametric study indicates that this orbit would have a short lifetime with the appropriate choice of initial apogee direction.

The apogee-burn option requires hardware changes to the upper stage; but the performance impact would be very small. The hardware changes include adding an apogee kick motor to the upper stage and some type of control jets to obtain the required burn attitude (because the spent upper stage is tumbled just after separation to ensure that it does not recontact the satellite). Only a little propellant would be required for the attitude-control burns (less than 10 lb for typical engine characteristics).

Concluding Remarks

The orbital-debris problem in the geosynchronous region is not well understood. The two analyses presented will help to define the problem, but much more analysis is required before policies can be defined. So far, the collision probabilities in the geosynchronous region appear to be small, but uncontrolled growth in the satellite population might create unacceptable collision probabilities. Techniques for controlling this environment are needed. Two such techniques, apogee and perigee burns, have been described here.

Acknowledgments

The author wishes to acknowledge the contribution of the following people: John Vedder for his assistance in the application of statistical theory to the collision problem, Panayiotis Fardelos for his assistance in performing the parametric study of orbit lifetime and in defining the techniques for controlling the environment, and Don Kessler for his invaluable guidance in completing this study.

References

[1]UN Committee on Peaceful Uses of Outer Space, "Physical Nature and Technical Attributes of the Geostationary Orbit," prepared by the A/AC. 105-203, Aug. 29, 1977; Addendum II, Dec. 1978.

[2]Mann, M. R. and Singpurwalla, N. D., "Extreme Value Distributions," SC5065.4TR, Rockwell International, 1980.

[3]Hoots, F. R. and Roehrich, R. L., "Models for Propagation of NORAD Element Sets," Aerospace Defense Command, Spacetrack Rept. No. 3, Dec. 1980.

[4]Stephens, M. A., "The Anderson-Darling Statistic," Tech. Rept. No. 39, Stanford Univ., Stanford, CA, Oct. 1979.

[5]Hechler, M. and Van der Ha, J. C., "The Probability of Collisions on the Geostationary Ring," Vol. 18, No. 4, *AIAA Journal,* July/Aug. 1981. pp 361–366.

[6]Chobotov, V. A., "Collision Hazard in Space," TR-0081 (6790)-1, The Aerospace Corp., El Segundo, CA, 1981.

[7]Mueller, A. C. and Graf, O. F., "A Study of the Lifetime of the Geosynchronous Transfer Satellites," ACM-TR-115, Analytical and Computational Mathematics, Inc., Houston, TX, 1979.

[8]McCormick, B. R., Fardelos, P. L., and Bond, V. R., "A Semi-Analytic Technique for Predicting Orbit Lifetime of Geosynchronous Satellites," AIAA Paper 86-2176-CP, Aug. 1986.

Prevention of Breakups: Upper-Stage Operational Procedures

Chapter 13

Current Operational Practice for the Delta Second Stage

Almon A. Gray Jr.*
NASA Goddard Space Flight Center, Greenbelt, Maryland

The Delta second stage (the 3910 stage, Fig. 1) has a propellant loading of 10,200 lb, including the oxidizer (nitrogen tetroxide) and fuel (aerozine-50, consisting of 50% hydrazine and 50% UDMH). Its single engine develops 9900-lb thrust. The propellant is pressure-fed by helium stored in two-24-in. titanium spheres at 4350 psi (design burst, 5500 psi) and regulated to 255 psi. If necessary, residual helium can be vented forward to separate the stage from the payload. The propellant tanks (54.7 in. in diam) carry 70 cu. ft each of fuel (forward) and oxidizer (aft). A common bulkhead separates the tanks. Tanks and bulkhead are made of 410 stainless steel with a design operating pressure of 300 psi (design burst, 375 psi).

The attitude during powered flight is controlled by gimballing the engine in pitch and yaw plus nitrogen roll jets. Attitude during coast is controlled by nitrogen jets in all axes. The nitrogen is supplied from a 16-in. titanium sphere at 4150 psi (design burst, 5760 psi), which is regulated to 255 psi. The regulated nitrogen also supplies the propellant settling jets used for zero-gravity restarts. Guidance and control is performed by a strapdown inertial sensor unit and computer with a preprogrammed flight sequence. Electrical power for control, telemetry, and engine systems is provided by 28-V Silvercel battery packs.

Two command destruct receivers (CDR) and two safe/arm devices (S/A) initiate opposite ends of a linear shaped charge on the tanks. Thermal protection for the vehicle consists of layers of aluminized Mylar with mesh separators. The tanks are covered with three layers and the engine compartment sides with five layers; the engine compartment bottom is also five layers, the outer layer being goldized Kapton.

At the end of the mission, 250 lb of propellants remain. The propellant-tank pressure is 200 psi, the helium sphere is vented to space, and the nitrogen sphere residual pressure is 1800 psi. The S/A's are armed, the CDR's are off, and the remaining battery life is 1–2 h.

---

*Engineering Development Manager, Delta Project.

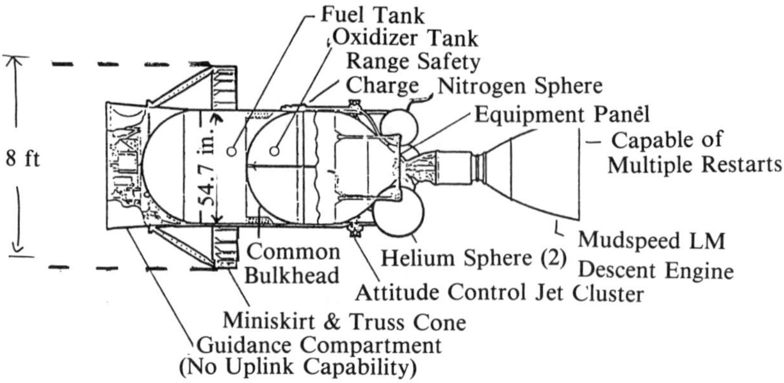

Fig. 1 Delta 3910 second stage. Dry mass, ≈ 1800 lb. Insulation blankets not shown.

After the Landsat-3 stage breakup in 1981, McDonnell Douglas Astronautics Company-West analyzed the probable cause and corrective measures to take. No single cause was found that fits all the data. Solar heating could cause overpressure of the oxidizer tank from helium plus vaporization of residual N_2O_4. However, one stage broke up before this could be a factor. Oxidizer-tank corrosion could weaken the tank enough to allow rupture at the residual tank pressure; but again, one stage broke up before this could be a factor. CDR's are known to turn back on after the guidance computer shuts down on low voltage and removes their shutoff command, but this can explain only the one early breakup.

The stages that broke up were compared with 13 stages on which experimental restarts had been conducted with the following results. All had been burned to at least sensed depletion, in which shutdown is commanded when acceleration falls below a preset threshold. All but one had oxidizer depletions (stage-loading calculations include a fuel bias). The one stage with higher-than-normal oxidizer residual was the only one that had broken up.

The major conclusion reached from the breakup analysis was that experimental restarts should be made standard, i.e., hold the engine on long enough to assure that as much oxidizer as possible is vented from the tank. Another conclusion was that the nitrogen sphere should be vented down, as analysis revealed a possibility of overpressure from heating if nitrogen usage is abnormally low.

In response to these conclusions, all (but trapped) propellant is vented by starting the engine and holding the propellant valves open for 1 min. On two-stage missions, a prior "evasive" burn and coast provides separation from spacecraft to preclude contamination. The tank pressures are reduced to about 75 psi by this venting. Nitrogen is vented by opening the propellant settling jets for 5 min, reducing sphere pressure essentially to zero.

None of the 12 experimental-restart stages that were intact in 1981 have since broken up. None of the 19 stages orbited since the above actions were taken have broken up.

Chapter 14

Ariane Third Stage: Current Operational Practice and Modifications Expected

R. Hergott*
CNES, Evry, France
and
E. Perez†
Arianespace, Evry, France

The Ariane third stage (Fig. 1; pipe and valve connections to hydrogen and oxygen tanks, Fig. 2) at main engine cut off (MECO), has a hydrogen-tank gas pressure of 2.7 bar and temperature of 150 K. The mass of liquid hydrogen remaining was 94 kg for Flight 16, and for all missions it has been less than 200 kg. Gas from the hydrogen tank feeds the attitude and roll control system (SCAR) with the maximum pressure controlled by safety relief valve 37 (see Fig. 2), which has a threshold at 3.15 bar. At MECO, the oxygen tank has a gas pressure of 2 bar and temperature of 170 K. The mass of liquid oxygen was 153 kg for Flight 16, and for all missions it has been less than 400 kg. When valve 34 (see Fig. 3) is open, gases are vented through the "Lox vents," which gives a longitudinal thrust. Valve 34 is both a commandable and passive safety relief valve with a threshold set to 2.35 bar. At MECO the helium pressurization tank has a gas pressure of 60–70 bar and a gas temperature of 135 K.

The vehicle has a command destruct system (Fig. 4) that uses linear destruct charges on the hydrogen and oxygen tanks. Activation of the destruct system will cut open both tanks.

For flight 16, Table 1 outlines the sequence of events after MECO. These maneuvers were needed to separate properly the spent third stage from the two spacecraft that formed the payload for the flight.

The functional and telemetry batteries run out of charge some 4 h after launch. The pyrotechnic batteries run out of charge about three months after launch. In Flights 01 to 12 and Flights 14 and 16, all valves were normally shut, except valve 42 and the chamber sweeping valve, which is normally open. When there is no more electric power, the valves open and the helium sphere exhausts into the chamber and into the hydrogen tank.

---

*Deputy to the Ariane 5 Program Director.

†Head of the Ariane Production Program Group.

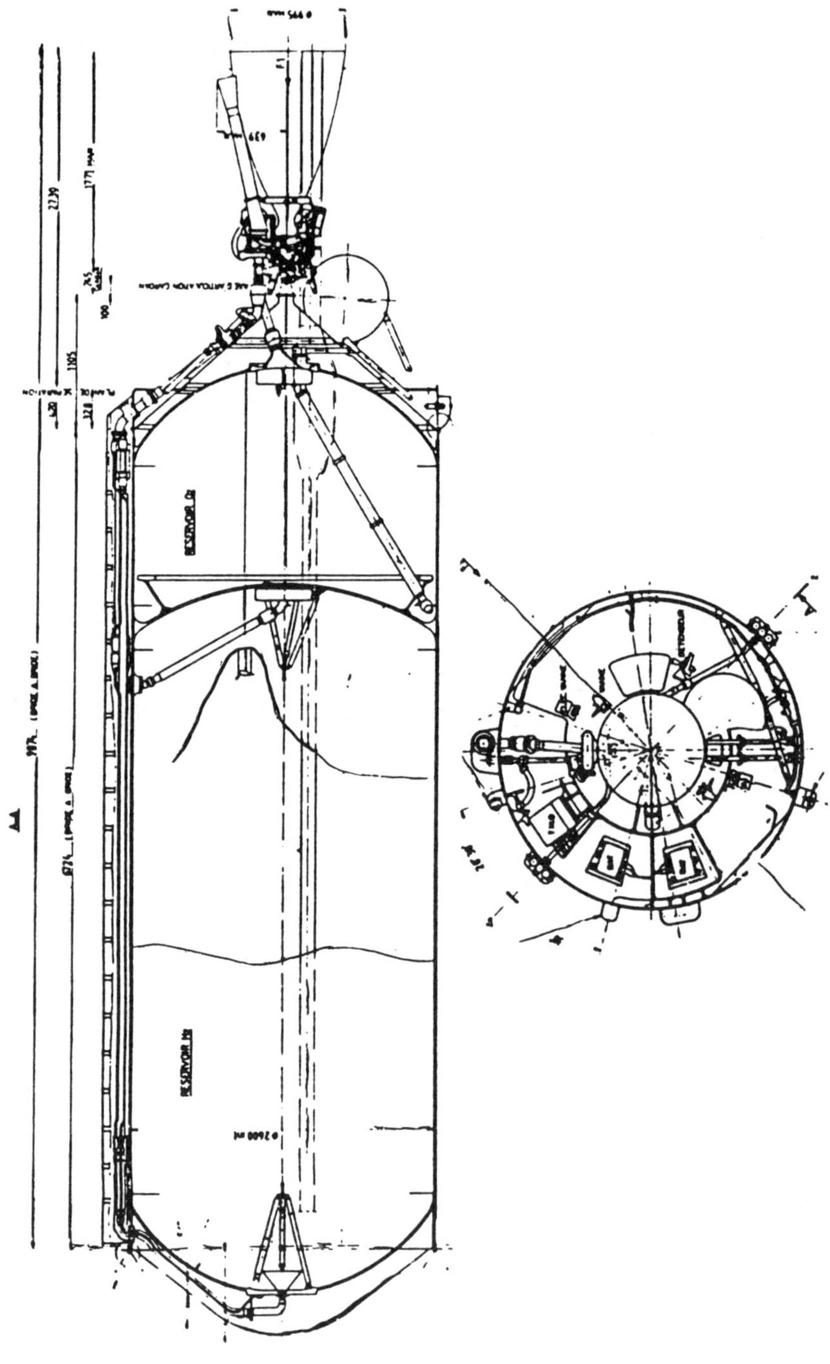

Fig. 1 Ariane third-stage views.

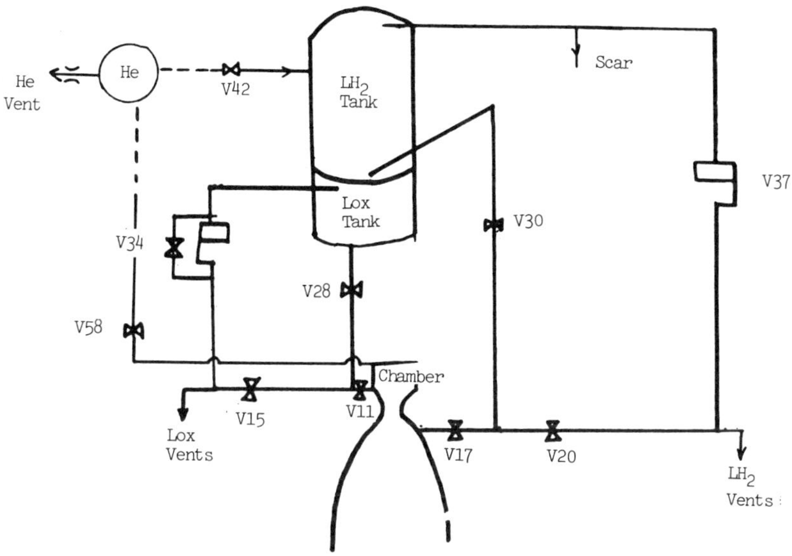

Fig. 2 Pipe and valve connections.

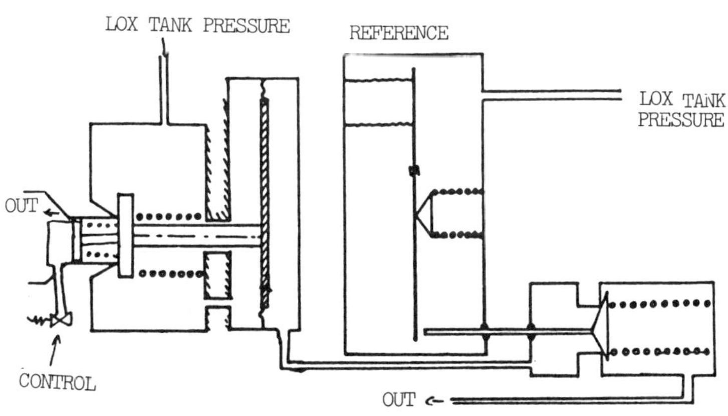

Fig. 3 Safety value 34 diagram.

Table 1 Flight 16 events after MECO

H3 = Ho + 844 S	Third stage engine cutoff (MECO)	Valves 11 and 17 shut down
Ho + 846 S	Start of SCAR control	Valve 34 (LOX vents) open
Ho + 870 S	Orientation of SPOT spacecraft	Valve 34 shut down
Ho + 877 S	Separation SPOT	
Ho + 880 S	Start of VIKING spacecraft orientation	
Ho + 909 S		Valve 34 open
Ho + 1066 S	Spin up to 10 rpm	
Ho + 1091 S	Separation of VIKING	
Ho + 1093 S		Valve 34 shut down
Ho + 1094 S	Despin of the 3rd stage	
Ho + 1115 S	Reorientation	
Ho + 1146 S		Valve 34 open
Ho + 1187 S	Spin up to 7 rpm	
H5 = Ho + 1244 S	End of mission (EOM)	All SCAR valves and valve 34 are shut down

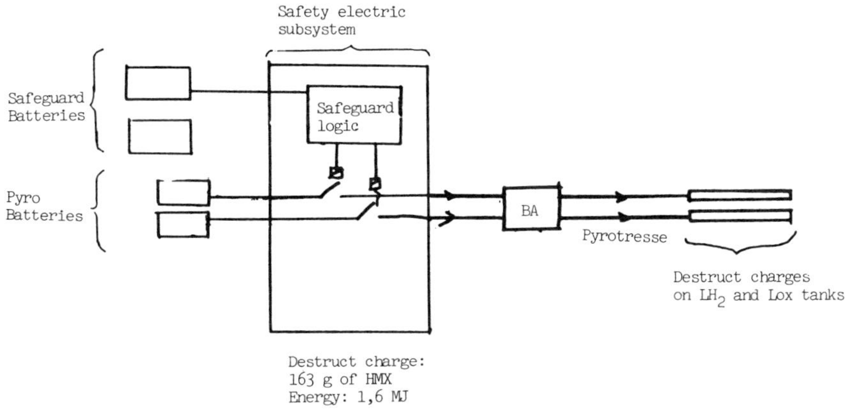

Fig. 4 Pyrotechnic circuit for destruction.

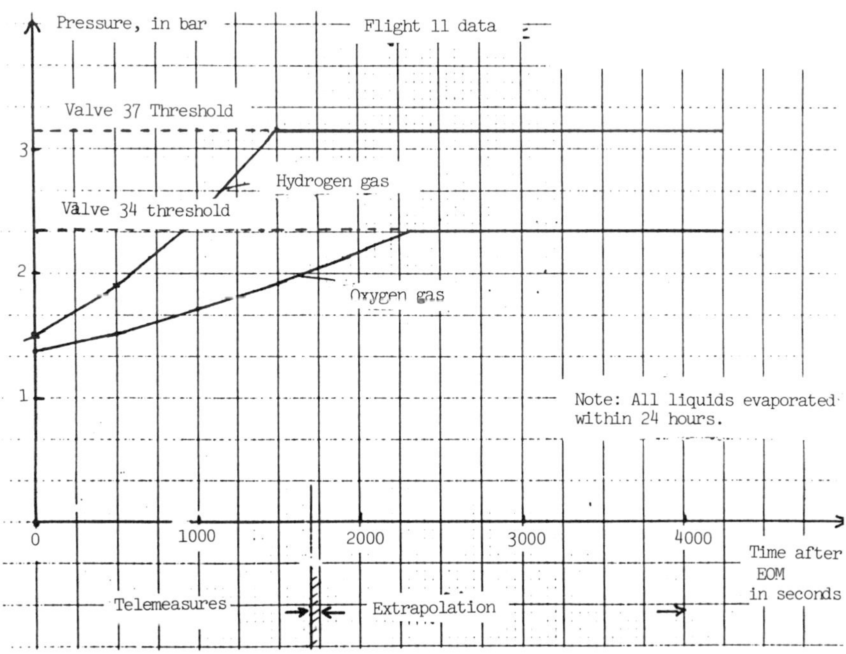

Fig. 5 Tank pressure history for Flight 11.

During flights 13 and 17 and future flights, all valves are normally shut. In all cases, the tanks are isolated at end of mission (EOM) and only the safety valves (34 and 37) keep the hydrogen and oxygen pressures below 3. 15 bar and 2.35 bar, respectively.

Figure 5 shows the evolution of pressure in the hydrogen and oxygen tanks for Flight 11. The pressure increases up to the value set by the relief-valve threshold and then remains constant. Calculations indicate that all the liquid in the tanks should evaporate within about 24 h.

The computed evolution of pressure in the helium tank, in both maximum and minimum solar flux (Fig. 6), drops below 40 bar in about 45 min after MECO.

Third stages in L01, V14, V16, and V17 all suffered breakups after launch (Table 2). The breakup of V16 has been studied most thoroughly. The following causes of third-stage breakup in Flight 16 have been examined and excluded from further consideration:

1) Explosion of an Ag Zn battery.
2) Accidental initiation of the destruct charges.
3) Spin due to hydrogen safety valve (37) exhaust thrust.
4) Electrostatic causes were ruled unlikely.

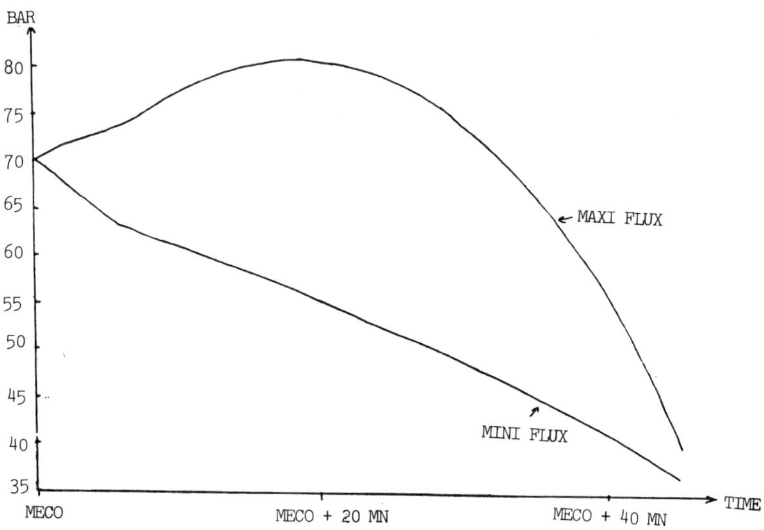

Fig. 6 Computed evolution of pressure in the helium tank for maximum and minimum solar flux.

Table 2 Breakup record of third stages

Vehicle	Orbit	Breakup	Orbit position
L01	GTO	90 days	—
L03 and L04	GTO	no data	—
L6 thru V13	GTO	no breakup	—
V14	GTO	48 hours	5th apogee
V16	HLS (SPOT)	10 months	In eclipse
V17	GTO	48 hours	5th apogee; out of eclipse

Structural failure due to gas pressure is a potential cause of breakup. If the following pressure limits for integrity of the tanks exceeded, structural failure was a possibility:

	Pressure,[a] bar	
	At 300 K	At 400 K
Hydrogen tank	≤ 3.9	3.4
Oxygen tank	≤ 3.4	2.9
Common bulkhead Δp	≤ 2.4	2.1
	≥ -0.3	2.1
Helium tank	≤ 240	205

[a]Equilibrium temperature is between 250 and 300 K.

Causes of the V16 breakup, and possibly also the other breakups, we conclude may be the following:

1) Bulkhead reversal due to leakage of the hydrogen tank.

2) Bulkhead rupture due to leakage of the oxygen tank.

3) Rupture of the hydrogen tank due to structural fatigue, after repeated expansions and contractions as the vehicle went in and out of eclipse.

To prevent breakups, corrective action needs to be taken to deplete hydrogen and oxygen liquids and reduce gas pressures as much as possible. This will be done for the next heliosynchronous mission: the launch of SPOT 2 in December 89 (Flight 36). It required hardware changes on valve 34 to the Lox vents, and on the hydrogen tank relief valve (37). With the adjunction of small pyrotechnic devices, these two valves can now be forced to a permanent passive state.

A sequence of events to get final stage depletion follows. After separation of the last payload, the third stage will be reoriented. Valve 34 will then be irreversibly forced open: after a proper time lag, sufficient to guarantee that the hydrogen pressure will always be higher than the oxygen pressure, valve 37 will also be irreversibly forced open. The new hardware has been qualified and the depletion sequence has been proved feasible. The third stage will be operated this way at least for every future heliosynchronous mission.

Chapter 15

Centaur Upper Stage

W. Groesbeck*
NASA Lewis Research Center, Cleveland, Ohio

The Centaur upper stage (Fig. 1, dry weight of 3800 lb) has pressure-stabilized, stainless-steel tanks 10 ft in diam and 30 ft long for liquid hydrogen (LH2) and liquid oxygen (LO2) under the following conditions: pressure ratings: max. operating pressure of LH2 tank = 30 psig, of LO2 tanks = 45 psig; proof pressure test for LH2 tank = 31 psig, for LO2 tanks = 52 psig; design ultimate of LH2 tank = 38.5 psig, of LO2 tanks = 64.6 psig. The skin thickness is 0.014 in. in the cylindrical section. The LO2 and LH2 tanks are separated by a double-walled, vacuum-insulated intermediate bulkhead. The LH2 tanks are protected with a layer of 1.0-in. thick, foam-filled fiberglass insulation panels during the ascent phase.

Propulsion in the main stage consists of two Pratt and Whitney RL10A 3-3A engines with a rated thrust of 33,000 lb, a specific impulse of 446.4 sec, and an 02/H2 mixture ratio of 5:1. The stage has autogenous pressurization of the LH2 tank and gaseous-helium pressurization of the LO2 tank.

Hamilton Standard monopropellant-hydrazine reaction-engine modules (REMs) make up the attitude-control system, with all-welded 304 stainless-steel components. There are two two-engine modules (6-lb thrust per engine) for lateral thrust and two four engine modules (6-lb thrust per engine) for lateral plus axial thrust. A titanium storage bottle for the hydrazine has an internal diaphragm with a nominal operating pressure of 450 psia.

The main-stage propellants are LH2 and LO2 with flight loading at liftoff of 30,750 lb. At the end of the mission the following residuals will be trapped in the system: LH2, 50 lb; gaseous H2, 25 lb; LO2, 65 lb; and gaseous O2, 170 lb. The RCS propellant, aniline-free anhydrous hydrazine with a maximum loading of 170 lb (one bottle), will have a residual at the end of the mission of 50 lb.

Pneumatics consist of a gaseous helium supply in two 26-in.-diam storage bottles made of titanium liners with a Kevlar wrap. These bottles have a charge pressure for flight of 3350 psia, proof-tested to 6966 psia. The residual at the end of the mission has a pressure of 700–1000 psia. There is

*Engineering Directorate.

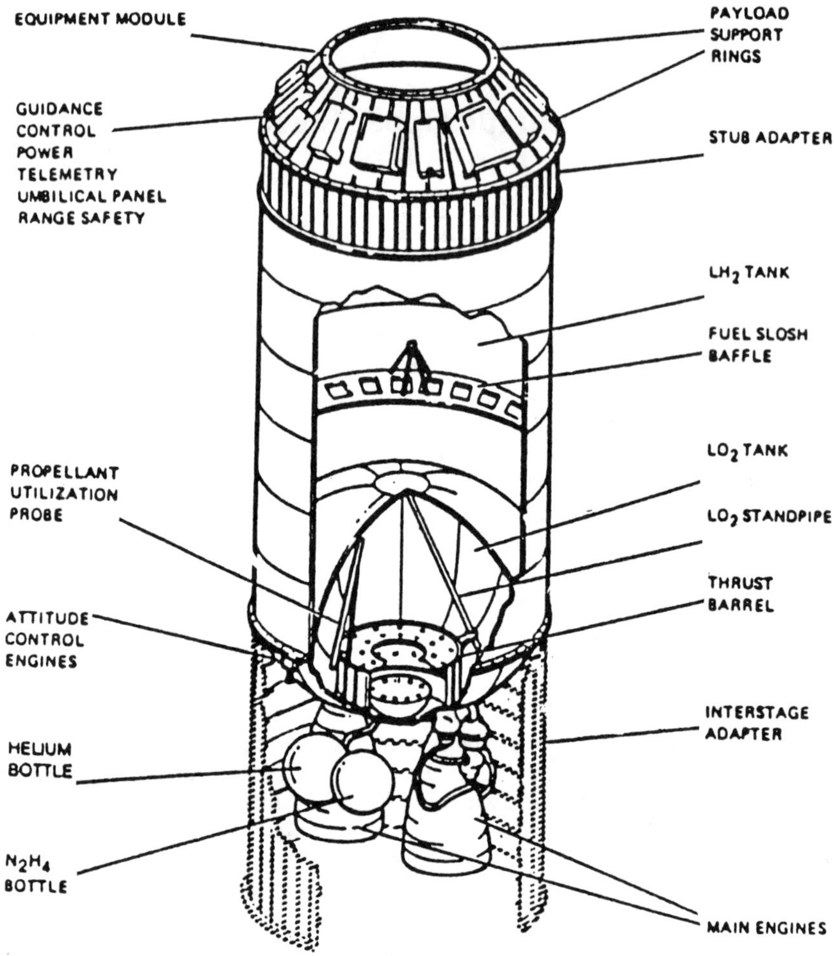

Fig. 1 Centaur vehicle arrangement.

a direct supply of the helium to the propellant-tank pressurization system. Also, regulated 450-psi gas is supplied to the vehicle's pneumatic system for actuation of engine control valves and pressurization of the hydrazine bottles and system.

Operations from Main Engine Cut Off (MECO) to spacecraft separation can be outlined as follows. At MECO plus 0.06 s, the RCS axial-thrust engines are commanded on, and they provide thrust to maintain propellants in settled position while venting down the LO2 and LH2 tanks to lower pressures. At MECO plus 2 s, the separation turn begins and the vehicle is reoriented to spacecraft separation vector. At MECO plus 30 s, the vehicle begins zero-*g* coast, the RCS engines are off, the tank pressures are vented and stable, but the venting is inhibited until after spacecraft separation. MECO plus 135 s, sees spacecraft separation and a following series of retro maneuvers.

At MECO plus 150 s, the Centaur is reoriented to the retro-maneuver vector preparatory to moving it out of sight of the spacecraft. At MECO plus 210 s, the RCS axial-thrust engines are turned on (4S on) and thrust applied to move the Centaur away from the spacecraft. This settles the propellants in preparation for propellant blowdown. AT MECO plus 255 s, the RCS axial thrust is reduced to 2S on, separation thrusting continues, and propellant retention is maintained. At MECO plus 330 s, the LH2 engine prestart valves are opened and the residual LH2 is blown down through the interstage cooldown valves. At MECO plus 331 s, the RCS axial thrust for propellant retention is terminated, and the vehicle attitude control is maintained with the RCS system. The LO2 prestart valves are opened at MECO plus 535 s, and the residual LO2 is blown down through the engines. At MECO plus 580 s, LH2 and LO2 tank blowdown is completed, and the LH2 and LO2 prestart valves are closed.

At MECO plus 3060 s, the RCS axial-thrust engines are turned on (4s on) to burn hydrazine to depletion (about 300 s needed). At MECO plus 4200 s, the LH2 and LO2 vent valves are unlocked, which puts the valves in a mechanical selfregulating mode. The LH2 tank pressure is maintained between 19 and 21.5 psia; the LO2 tank, between 29 and 32 psia. All other systems are left powered up until depletion of the electrical power. At MECO plus 5640 s, the Centaur vehicle should have entirely lost electrical power, leaving it inert and drifting in orbit.

The planetary launch program (Moon, Mercury, Venus, Mars) during 1966–78 entailed 23 launches but no known re-entries. The low-latitude Earth orbit missions (HEAO, OAO) during 1963–79 involved nine launches and five known re-entries, the last in 1979. The geosynchronous missions (Intelsat, Fltsatcom, Comstar) from 1969–86 saw 32 launches but only three known re-entries (the last in 1983). There have been no known breakups of the Centaur upper-stage vehicle.

Chapter 16

Safing of H-1 Second Stage After Spacecraft Separation

K. Noda*
National Space Development Agency of Japan, Tokyo, Japan

To prevent the explosion of expended orbiting rocket stages, it is necessary to safe them at the end of a mission. In addition to the existing measures for this, we plan a propellant-depletion burn of the H-1 second stage by an idle-mode firing of its liquid-oxygen (LOX)/liquid-hydrogen (LH2) LE-5 engine.

Figure 1 shows a schematic diagram of the H-1 second stage. The overall propellant tank is divided into two separate tanks by a common bulkhead. The lower tank, containing LOX, is controlled at a higher pressure than the upper, LH2 tank in order to maintain a positive pressure differential across the bulkhead.

At the end of the mission, the stage contains approximately 1000 liters of LH2 and 200 liters of LOX. This reserve gets held to ensure a commanded (rather than a depletion) shutdown of the engine. Some reserve propellant will be blown down thorough engine chill-down valves (FCV, LCV), but most of it will remain in the tanks, there to evaporate and be vented through mechanical, selfregulating relief valves (FVRV, OVRV). This keeps the pressures of each tank and pressure differential across the common bulkhead within structural limits.

These measures were performed successfully in the two flights of H-1 engineering test vehicle. However, it takes a long time for all the residual propellant to evaporate entirely. In the first H-1 test flight, it was observed that the expended second stage was rotating slowly in orbit, and displayed a white gaseous discharge assumed to originate from residual propellant.

The second stage will be safer against breakup from overpressure or explosion if the residual propellant can be blown down faster. An idle-mode firing of the LE-5 should serve this purpose.

Figure 2 shows the schematic diagram of the LE-5 engine and its normal operational sequence. LE-5 is a gas generator cycle engine with a unique method for start-up (''coolant bleed start''), and is equipped with an engine

Invited paper.

*Associate Senior Engineer, Launch Vehicle Group.

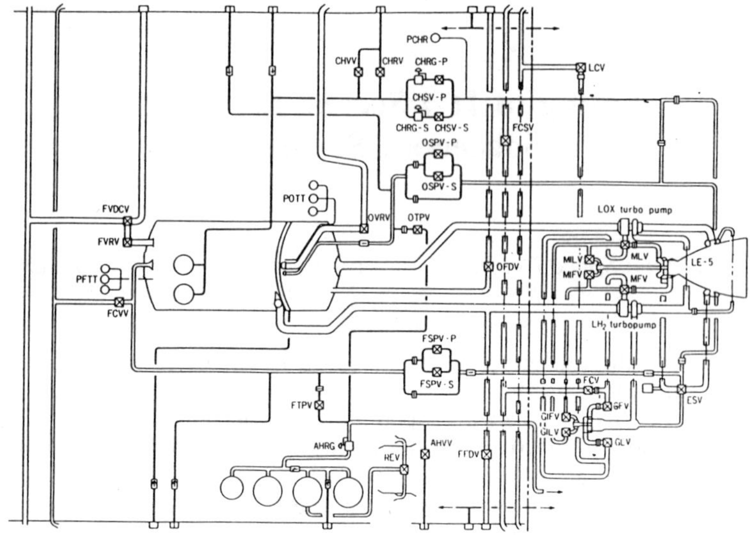

AHRG	Ambient Herium Regurator
AHVV	Ambient Herium Vent Valve
CHRG	Cryogenic Herium Regurator
CHRV	Cryogenic Herium Relief Valve
CHSV	Cryogenic Herium Shut-off Valve
CHVV	Cryogenic Herium Vent Valve
FCSV	LH2 Chill-down line Shut-off Valve
FCVV	LH2 Coast Vent Valve
FFDV	LH2 Fill and Drain Valve
FSPV	LH2 tank Steady-state Pressurization Valve
FTPV	LH2 tank Transient Pressurization Valve
FVDCV	LH2 tank Vent Direction Control Valve
FVRV	LH2 tank Vent and Relief Valve
OFDV	LOX Fill and Drain Valve
OSPV	LOX tank Steady-state Pressurization Valve
OTPV	LOX tank Transient Pressurization Valve
OVRV	LOX tank Vent and Relief Valve
PFTT	Pressure transducer, LH2 Tank Top
POTT	Pressure transducer, LOX Tank Top
REV	Retro system Explosive Valve

Fig. 1 H-1 second stage schematic.

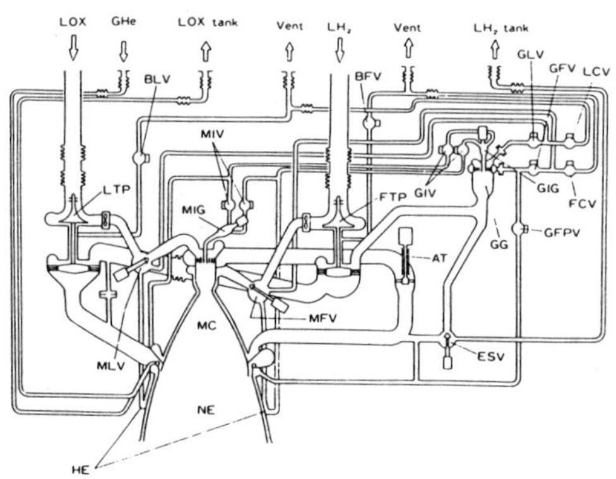

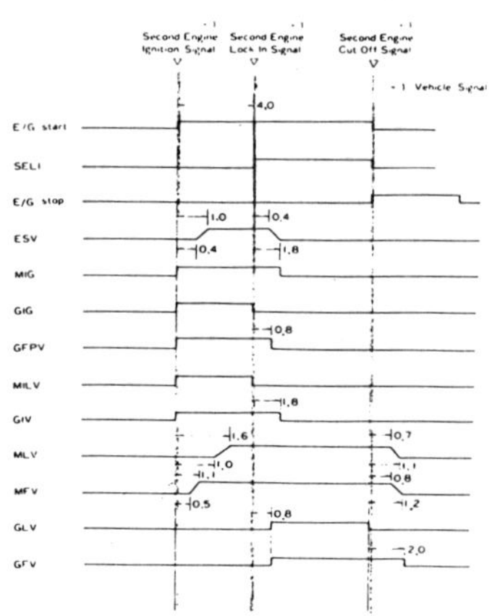

AT	Auxiliary Turbine
ESV	Engine Start Valve
FCV	Fuel Chill-down Valve
FTP	Fuel Turbopump
GG	Gas Generator
GFV	GG Fuel Valve
GFPV	GG Fuel Purge Valve
GIG	GG Ignitor
GIV	GG Ignitor Valve
GLV	GG LOX Valve
HE	Heat Exchanger
LTP	LOX Turbopump
MC	Main Combustion Chamber
MFV	Main Fuel Valve
MIG	MC Ignitor
MIV	MIG Valve
MLV	Main LOX Valve
NE	Nozzle Extension
BFV	FTP Bearing Valve
BLV	LTP Bearing Valve

Fig. 2 LE-5 rocket engine schematic.

start valve (ESV) to tap off gaseous fuel from the regenerative cooling jacket of the main combustion chamber to the turbopump turbines.

For a normal firing of the engine, a start signal is received from the second stage, ignitors are activated, and the main propellant valves (MFV, MLV) and ESV are opened. The engine starts under a tank pressure-fed condition, then heated hydrogen from the regeneration cooling passages is tapped off to the turbopump turbines and drives them. Thus, the engine thrust builds up automatically. When 50% of the rated thrust is reached, the engine closes ESV, activates the gas generator (GG) simultaneously, and advances to a steady GG cycle.

For idle mode firing, the engine starts without opening ESV and does not use the turbopumps. It operates under a tank pressure-fed condition only, and does not advance to the gas generator cycle. It is possible to blow down the residual propellants within a few hundreds of seconds by the idle mode firing. It requires no change in the engine hardware, except for minor modifications of the engine control box.

The H-11 rocket, which will succeed the H-1 rocket, will use an improved LE-5 engine (designated LE-5A). The idle mode firing capability is being studied for LE-5A not only to deplete the residual propellants, but also to utilize the loaded propellants more effectively.

It is difficult to carry out an accurate ground test of the idle mode firing capability, because the required engine chamber pressure is very low. In this sense, the idle mode firing of the LE-5 engine in the H-1 vehicle is also an engineering test to get data for the LE-5A engine.

Author Index

PROGRESS IN ASTRONAUTICS AND AERONAUTICS SERIES VOLUMES

VOLUME TITLE/EDITORS

***1. Solid Propellant Rocket Research** (1960)
Martin Summerfield
Princeton University

***2. Liquid Rockets and Propellants** (1960)
Loren E. Bollinger
The Ohio State University
Martin Goldsmith
The Rand Corporation
Alexis W. Lemmon Jr.
Battelle Memorial Institute

***3. Energy Conversion for Space Power** (1961)
Nathan W. Snyder
Institute for Defense Analyses

***4. Space Power Systems** (1961)
Nathan W. Snyder
Institute for Defense Analyses

***5. Electrostatic Propulsion** (1961)
David B. Langmuir
Space Technology Laboratories, Inc.
Ernst Stuhlinger
NASA George C. Marshall Space Flight Center
J.M. Sellen Jr.
Space Technology Laboratories, Inc.

***6. Detonation and Two-Phase Flow** (1962)
S.S. Penner
California Institute of Technology
F.A. Williams
Harvard University

***7. Hypersonic Flow Research** (1962)
Frederick R. Riddell
AVCO Corporation

***8. Guidance and Control** (1962)
Robert E. Roberson
Consultant
James S. Farrior
Lockheed Missiles and Space Company

***9. Electric Propulsion Development** (1963)
Ernst Stuhlinger
NASA George C. Marshall Space Flight Center

***10. Technology of Lunar Exploration** (1963)
Clifford I. Cummings
Harold R. Lawrence
Jet Propulsion Laboratory

***11. Power Systems for Space Flight** (1963)
Morris A. Zipkin
Russell N. Edwards
General Electric Company

12. Ionization in High-Temperature Gases (1963)
Kurt E. Shuler, Editor
National Bureau of Standards
John B. Fenn, Associate Editor
Princeton University

***13. Guidance and Control—II** (1964)
Robert C. Langford
General Precision Inc.
Charles J. Mundo
Institute of Naval Studies

***14. Celestial Mechanics and Astrodynamics** (1964)
Victor G. Szebehely
Yale University Observatory

***15. Heterogeneous Combustion** (1964)
Hans G. Wolfhard
Institute for Defense Analyses
Irvin Glassman
Princeton University
Leon Green Jr.
Air Force Systems Command

16. Space Power Systems Engineering (1966)
George C. Szego
Institute for Defense Analyses
J. Edward Taylor
TRW Inc.

17. Methods in Astrodynamics and Celestial Mechanics (1966)
Raynor L. Duncombe
U.S. Naval Observatory
Victor G. Szebehely
Yale University Observatory

18. Thermophysics and Temperature Control of Spacecraft and Entry Vehicles (1966)
Gerhard B. Heller
NASA George C. Marshall Space Flight Center

***19. Communication Satellite Systems Technology** (1966)
Richard B. Marsten
Radio Corporation of America

*Out of print.

20. Thermophysics of Spacecraft and Planetary Bodies: Radiation Properties of Solids and the Electromagnetic Radiation Environment in Space (1967)
Gerhard B. Heller
NASA George C. Marshall Space Flight Center

21. Thermal Design Principles of Spacecraft and Entry Bodies (1969)
Jerry T. Bevans
TRW Systems

22. Stratospheric Circulation (1969)
Willis L. Webb
Atmospheric Sciences Laboratory, White Sands, and University of Texas at El Paso

23. Thermophysics: Applications to Thermal Design of Spacecraft (1970)
Jerry T. Bevans
TRW Systems

24. Heat Transfer and Spacecraft Thermal Control (1971)
John W. Lucas
Jet Propulsion Laboratory

25. Communication Satellites for the 70's: Technology (1971)
Nathaniel E. Feldman
The Rand Corporation
Charles M. Kelly
The Aerospace Corporation

26. Communication Satellites for the 70's: Systems (1971)
Nathaniel E. Feldman
The Rand Corporation
Charles M. Kelly
The Aerospace Corporation

27. Thermospheric Circulation (1972)
Willis L. Webb
Atmospheric Sciences Laboratory, White Sands, and University of Texas at El Paso

28. Thermal Characteristics of the Moon (1972)
John W. Lucas
Jet Propulsion Laboratory

29. Fundamentals of Spacecraft Thermal Design (1972)
John W. Lucas
Jet Propulsion Laboratory

30. Solar Activity Observations and Predictions (1972)
Patrick S. McIntosh
Murray Dryer
Environmental Research Laboratories, National Oceanic and Atmospheric Administration

31. Thermal Control and Radiation (1973)
Chang-Lin Tien
University of California at Berkeley

32. Communications Satellite Systems (1974)
P.L. Bargellini
COMSAT Laboratories

33. Communications Satellite Technology (1974)
P.L. Bargellini
COMSAT Laboratories

34. Instrumentation for Airbreathing Propulsion (1974)
Allen E. Fuhs
Naval Postgraduate School
Marshall Kingery
Arnold Engineering Development Center

35. Thermophysics and Spacecraft Thermal Control (1974)
Robert G. Hering
University of Iowa

36. Thermal Pollution Analysis (1975)
Joseph A. Schetz
Virginia Polytechnic Institute

37. Aeroacoustics: Jet and Combustion Noise; Duct Acoustics (1975)
Henry T. Nagamatsu, Editor
General Electric Research and Development Center
Jack V. O'Keefe, Associate Editor
The Boeing Company
Ira R. Schwartz, Associate Editor
NASA Ames Research Center

38. Aeroacoustics: Fan, STOL, and Boundary Layer Noise; Sonic Boom; Aeroacoustic Instrumentation (1975)
Henry T. Nagamatsu, Editor
General Electric Research and Development Center
Jack V. O'Keefe, Associate Editor
The Boeing Company
Ira R. Schwartz, Associate Editor
NASA Ames Research Center

39. Heat Transfer with Thermal Control Applications (1975)
M. Michael Yovanovich
University of Waterloo

40. Aerodynamics of Base Combustion (1976)
S.N.B. Murthy, Editor
Purdue University
J.R. Osborn, Associate Editor
Purdue University
A.W. Barrows
J.R. Ward, Associate Editors
Ballistics Research Laboratories

41. Communications Satellite Developments: Systems (1976)
Gilbert E. LaVean
Defense Communications Agency
William G. Schmidt
CML Satellite Corporation

42. Communications Satellite Developments: Technology (1976)
William G. Schmidt
CML Satellite Corporation
Gilbert E. LaVean
Defense Communications Agency

43. Aeroacoustics: Jet Noise, Combustion and Core Engine Noise (1976)
Ira R. Schwartz, Editor
NASA Ames Research Center
Henry T. Nagamatsu, Associate Editor
General Electric Research and Development Center
Warren C. Strahle, Associate Editor
Georgia Institute of Technology

44. Aeroacoustics: Fan Noise and Control; Duct Acoustics; Rotor Noise (1976)
Ira R. Schwartz, Editor
NASA Ames Research Center
Henry T. Nagamatsu, Associate Editor
General Electric Research and Development Center
Warren C. Strahle, Associate Editor
Georgia Institute of Technology

45. Aeroacoustics: STOL Noise; Airframe and Airfoil Noise (1976)
Ira R. Schwartz, Editor
NASA Ames Research Center
Henry T. Nagamatsu, Associate Editor
General Electric Research and Development Center
Warren C. Strahle, Associate Editor
Georgia Institute of Technology

46. Aeroacoustics: Acoustic Wave Propagation; Aircraft Noise Prediction; Aeroacoustic Instrumentation (1976)
Ira R. Schwartz, Editor
NASA Ames Research Center
Henry T. Nagamatsu, Associate Editor
General Electric Research and Development Center
Warren C. Strahle, Associate Editor
Georgia Institute of Technology

47. Spacecraft Charging by Magnetospheric Plasmas (1976)
Alan Rosen
TRW Inc.

48. Scientific Investigations on the Skylab Satellite (1976)
Marion I. Kent
Ernst Stuhlinger
NASA George C. Marshall Space Flight Center
Shi-Tsan Wu
The University of Alabama

49. Radiative Transfer and Thermal Control (1976)
Allie M. Smith
ARO Inc.

50. Exploration of the Outer Solar System (1976)
Eugene W. Greenstadt
TRW Inc.
Murray Dryer
National Oceanic and Atmospheric Administration
Devrie S. Intriligator
University of Southern California

51. Rarefied Gas Dynamics, Parts I and II (two volumes) (1977)
J. Leith Potter
ARO Inc.

52. Materials Sciences in Space with Application to Space Processing (1977)
Leo Steg
General Electric Company

53. Experimental Diagnostics in Gas Phase Combustion Systems (1977)
Ben T. Zinn, Editor
Georgia Institute of Technology
Craig T. Bowman, Associate Editor
Stanford University
Daniel L. Hartley, Associate Editor
Sandia Laboratories
Edward W. Price, Associate Editor
Georgia Institute of Technology
James G. Skifstad, Associate Editor
Purdue University

54. Satellite Communications: Future Systems (1977)
David Jarett
TRW Inc.

55. Satellite Communications: Advanced Technologies (1977)
David Jarett
TRW Inc.

56. Thermophysics of Spacecraft and Outer Planet Entry Probes (1977)
Allie M. Smith
ARO Inc.

57. Space-Based Manufacturing from Nonterrestrial Materials (1977)
Gerard K. O'Neill, Editor
Princeton University
Brian O'Leary, Assistant Editor
Princeton University

58. Turbulent Combustion (1978)
Lawrence A. Kennedy
State University of New York at Buffalo

59. Aerodynamic Heating and Thermal Protection Systems (1978)
Leroy S. Fletcher
University of Virginia

60. Heat Transfer and Thermal Control Systems (1978)
Leroy S. Fletcher
University of Virginia

61. Radiation Energy Conversion in Space (1978)
Kenneth W. Billman
NASA Ames Research Center

62. Alternative Hydrocarbon Fuels: Combustion and Chemical Kinetics (1978)
Craig T. Bowman
Stanford University
Jorgen Birkeland
Department of Energy

63. Experimental Diagnostics in Combustion of Solids (1978)
Thomas L. Boggs
Naval Weapons Center
Ben T. Zinn
Georgia Institute of Technology

64. Outer Planet Entry Heating and Thermal Protection (1979)
Raymond Viskanta
Purdue University

65. Thermophysics and Thermal Control (1979)
Raymond Viskanta
Purdue University

66. Interior Ballistics of Guns (1979)
Herman Krier
University of Illinois at Urbana-Champaign
Martin Summerfield
New York University

***67. Remote Sensing of Earth from Space: Role of "Smart Sensors"** (1979)
Roger A. Breckenridge
NASA Langley Research Center

68. Injection and Mixing in Turbulent Flow (1980)
Joseph A. Schetz
Virginia Polytechnic Institute and State University

69. Entry Heating and Thermal Protection (1980)
Walter B. Olstad
NASA Headquarters

70. Heat Transfer, Thermal Control, and Heat Pipes (1980)
Walter B. Olstad
NASA Headquarters

71. Space Systems and Their Interactions with Earth's Space Environment (1980)
Henry B. Garrett
Charles P. Pike
Hanscom Air Force Base

72. Viscous Flow Drag Reduction (1980)
Gary R. Hough
Vought Advanced Technology Center

73. Combustion Experiments in a Zero-Gravity Laboratory (1981)
Thomas H. Cochran
NASA Lewis Research Center

74. Rarefied Gas Dynamics, Parts I and II (two volumes) (1981)
Sam S. Fisher
University of Virginia at Charlottesville

75. Gasdynamics of Detonations and Explosions (1981)
J.R. Bowen
University of Wisconsin at Madison
N. Manson
Université de Poitiers
A.K. Oppenheim
University of California at Berkeley
R.I. Soloukhin
Institute of Heat and Mass Transfer, BSSR Academy of Sciences

76. Combustion in Reactive Systems (1981)
J.R. Bowen
University of Wisconsin at Madison
N. Manson
Université de Poitiers
A.K. Oppenheim
University of California at Berkeley
R.I. Soloukhin
Institute of Heat and Mass Transfer, BSSR Academy of Sciences

77. Aerothermodynamics and Planetary Entry (1981)
A.L. Crosbie
University of Missouri-Rolla

78. Heat Transfer and Thermal Control (1981)
A.L. Crosbie
University of Missouri-Rolla

79. Electric Propulsion and Its Applications to Space Missions (1981)
Robert C. Finke
NASA Lewis Research Center

80. Aero-Optical Phenomena (1982)
Keith G. Gilbert
Leonard J. Otten
Air Force Weapons Laboratory

81. Transonic Aerodynamics (1982)
David Nixon
Nielsen Engineering & Research, Inc.

82. Thermophysics of Atmospheric Entry (1982)
T.E. Horton
The University of Mississippi

83. Spacecraft Radiative Transfer and Temperature Control (1982)
T.E. Horton
The University of Mississippi

84. Liquid-Metal Flows and Magnetohydrodynamics (1983)
H. Branover
Ben-Gurion University of the Negev
P.S. Lykoudis
Purdue University
A. Yakhot
Ben-Gurion University of the Negev

85. Entry Vehicle Heating and Thermal Protection Systems: Space Shuttle, Solar Starprobe, Jupiter Galileo Probe (1983)
Paul E. Bauer
McDonnell Douglas Astronautics Company
Howard E. Collicott
The Boeing Company

86. Spacecraft Thermal Control, Design, and Operation (1983)
Howard E. Collicott
The Boeing Company
Paul E. Bauer
McDonnell Douglas Astronautics Company

87. Shock Waves, Explosions, and Detonations (1983)
J.R. Bowen
University of Washington
N. Manson
Université de Poitiers
A.K. Oppenheim
University of California at Berkeley
R.I. Soloukhin
Institute of Heat and Mass Transfer, BSSR Academy of Sciences

88. Flames, Lasers, and Reactive Systems (1983)
J.R. Bowen
University of Washington
N. Manson
Université de Poitiers
A.K. Oppenheim
University of California at Berkeley
R.I. Soloukhin
Institute of Heat and Mass Transfer, BSSR Academy of Sciences

89. Orbit-Raising and Maneuvering Propulsion: Research Status and Needs (1984)
Leonard H. Caveny
Air Force Office of Scientific Research

90. Fundamentals of Solid-Propellant Combustion (1984)
Kenneth K. Kuo
The Pennsylvania State University
Martin Summerfield
Princeton Combustion Research Laboratories, Inc.

91. Spacecraft Contamination: Sources and Prevention (1984)
J.A. Roux
The University of Mississippi
T.D. McCay
NASA Marshall Space Flight Center

92. Combustion Diagnostics by Nonintrusive Methods (1984)
T.D. McCay
NASA Marshall Space Flight Center
J.A. Roux
The University of Mississippi

93. The INTELSAT Global Satellite System (1984)
Joel Alper
COMSAT Corporation
Joseph Pelton
INTELSAT

94. Dynamics of Shock Waves, Explosions, and Detonations (1984)
J.R. Bowen
University of Washington
N. Manson
Université de Poitiers
A.K. Oppenheim
University of California
R.I. Soloukhin
Institute of Heat and Mass Transfer, BSSR Academy of Sciences

95. Dynamics of Flames and Reactive Systems (1984)
J.R. Bowen
University of Washington
N. Manson
Université de Poitiers
A.K. Oppenheim
University of California
R.I. Soloukhin
Institute of Heat and Mass Transfer, BSSR Academy of Sciences

96. Thermal Design of Aeroassisted Orbital Transfer Vehicles (1985)
H.F. Nelson
University of Missouri-Rolla

97. Monitoring Earth's Ocean, Land, and Atmosphere from Space – Sensors, Systems, and Applications (1985)
Abraham Schnapf
Aerospace Systems Engineering

98. Thrust and Drag: Its Prediction and Verification (1985)
Eugene E. Covert
Massachusetts Institute of Technology
C.R. James
Vought Corporation
William F. Kimzey
Sverdrup Technology AEDC Group
George K. Richey
U.S. Air Force
Eugene C. Rooney
U.S. Navy Department of Defense

99. Space Stations and Space Platforms — Concepts, Design, Infrastructure, and Uses (1985)
Ivan Bekey
Daniel Herman
NASA Headquarters

100. Single- and Multi-Phase Flows in an Electromagnetic Field Energy, Metallurgical, and Solar Applications (1985)
Herman Branover
Ben-Gurion University of the Negev
Paul S. Lykoudis
Purdue University
Michael Mond
Ben-Gurion University of the Negev

101. MHD Energy Conversion: Physiotechnical Problems (1986)
V.A. Kirillin
A.E. Sheyndlin
Soviet Academy of Sciences

102. Numerical Methods for Engine-Airframe Integration (1986)
S.N.B. Murthy
Purdue University
Gerald C. Paynter
Boeing Airplane Company

103. Thermophysical Aspects of Re-Entry Flows (1986)
James N. Moss
NASA Langley Research Center
Carl D. Scott
NASA Johnson Space Center

104. Tactical Missile Aerodynamics (1986)
M.J. Hemsch
PRC Kentron, Inc.
J.N. Nielsen
NASA Ames Research Center

105. Dynamics of Reactive Systems Part I: Flames and Configurations; Part II: Modeling and Heterogeneous Combustion (1988)
J.R. Bowen
University of Washington
J.-C. Leyer
Université de Poitiers
R.I. Soloukhin
Institute of Heat and Mass Transfer, BSSR Academy of Sciences

106. Dynamics of Explosions (1986)
J.R. Bowen
University of Washington
J.-C. Leyer
Université de Poitiers
R.I. Soloukhin
Institute of Heat and Mass Transfer, BSSR Academy of Sciences

107. Spacecraft Dielectric Material Properties and Spacecraft Charging (1986)
A.R. Frederickson
U.S. Air Force Rome Air Development Center
D.B. Cotts
SRI International
J.A. Wall
U.S. Air Force Rome Air Development Center
F.L. Bouquet
Jet Propulsion Laboratory, California Institute of Technology

108. Opportunities for Academic Research in a Low-Gravity Environment (1986)
George A. Hazelrigg
National Science Foundation
Joseph M. Reynolds
Louisiana State University

109. Gun Propulsion Technology (1988)
Ludwig Stiefel
U.S. Army Armament Research, Development and Engineering Center

110. Commercial Opportunities in Space (1988)
F. Shahrokhi
K.E. Harwell
University of Tennessee Space Institute
C.C. Chao
National Cheng Kung University

111. Liquid-Metal Flows: Magnetohydrodynamics and Applications (1988)
Herman Branover, Michael Mond, and Yeshajahu Unger
Ben-Gurion University of the Negev

112. Current Trends in Turbulence Research (1988)
Herman Branover, Michael Mond, and Yeshajahu Unger
Ben-Gurion University of the Negev

113. Dynamics of Reactive Systems Part I: Flames; Part II: Heterogeneous Combustion and Applications (1988)
A.L. Kuhl
R & D Associates
J.R. Bowen
University of Washington
J.-C. Leyer
Université de Poitiers
A. Borisov
USSR Academy of Sciences

114. Dynamics of Explosions (1988)
A.L. Kuhl
R & D Associates
J.R. Bowen
University of Washington
J.-C. Leyer
Université de Poitiers
A. Borisov
USSR Academy of Sciences

115. Machine Intelligence and Autonomy for Aerospace (1988)
E. Heer
Heer Associates, Inc.
H. Lum
NASA Ames Research Center

116. Rarefied Gas Dynamics: Space-Related Studies (1989)
E.P. Muntz
University of Southern California
D.P. Weaver
U.S. Air Force Astronautics Laboratory (AFSC)
D.H. Campbell
The University of Dayton Research Institute

117. Rarefied Gas Dynamics: Physical Phenomena (1989)
E.P. Muntz
University of Southern California
D.P. Weaver
U.S. Air Force Astronautics Laboratory (AFSC)
D.H. Campbell
The University of Dayton Research Institute

118. Rarefied Gas Dynamics: Theoretical and Computational Techniques (1989)
E.P. Muntz
University of Southern California
D.P. Weaver
U.S. Air Force Astronautics Laboratory (AFSC)
D.H. Campbell
The University of Dayton Research Institute

119. Test and Evaluation of the Tactical Missile (1989)
Emil J. Eichblatt Jr.
Pacific Missile Test Center

120. Unsteady Transonic Aerodynamics (1989)
David Nixon
Nielsen Engineering & Research, Inc.

121. Orbital Debris from Upper-Stage Breakup (1989)
Joseph P. Loftus Jr.
NASA Johnson Space Center

(Other Volumes are planned.)